MAMMALS
OF THE WORLD

D0120723

In the same series

BIRDS OF THE WORLD
Hans Hvass

FISHES OF THE WORLD
Hans Hvass

REPTILES AND AMPHIBIANS
OF THE WORLD
Hans Hvass

PREHISTORIC LIFE ON EARTH
Kai Petersen

MAMMALS OF THE WORLD

Hans Hvass

Translated by
Gwynne Vevers

Illustrated by
Wilhelm Eigener

A METHUEN PAPERBACK
EYRE METHUEN
LONDON

First published by Politikens Forlag in 1956
as Alverdens Pattedyr

Copyright in all countries signatory to the Berne Convention
English Translation first published 1961
Methuen Paperback edition published 1975

© *1961 by Methuen & Co Ltd*

Printed for Eyre Methuen Ltd. 11 New Fetter Lane, London EC4P
4EE by Aarhuus Stiftsbogtrykkerie A/S

ISBN 0 413 34140 2

This book is sold subject to the condition that it shall not, by
way of trade or otherwise, be lent, resold, hired out, or otherwise
circulated without the publisher's prior consent in any form of
binding or cover other than that in which it is published and
without a similar condition including this condition being imposed
on the subsequent purchaser.

Contents

Foreword

THIS book deals with the mammals of the world, but not with all the mammals of the world—this would be impossible in a book of this size, for there are about 3,200 mammals living in the world today. In selecting the mammals depicted, emphasis has been placed not only on the common species but also on the rarer animals, many of which show interesting adaptations to special environments or are now very restricted in distribution.

Several methods of arranging or classifying the different groups of mammals have been suggested by zoologists and there is still no general agreement. In the present book we have started with the Primates (apes, monkeys and lemurs) and finished with the Marsupials and the primitive egg-laying Monotremes. It should, however, be mentioned that in many modern classifications, and in particular in that of G. G. Simpson, the arrangement is almost the other way round, and starts with the primitive mammals and finishes with the more highly evolved. However, the arguments on which

these classifications are based are highly technical and need not concern those who are primarily interested in learning about the great range of mammals still to be found in the world of today.

The mammals are just one class within the group known as the Vertebrates, or animals with backbones. The other classes of Vertebrates are the birds, reptiles, amphibians, bony fishes, cartilaginous fishes (skates, dogfish and sharks) and cyclostomes (lampreys and hagfish). The mammals themselves are divided into seventeen orders: Primates, Ungulates, Elephants, Sirenians, Hyraxes, Whales, Seals, Carnivores, Rodents, Edentates, Pangolins, Aardvarks, Bats, Insectivores, Flying Lemurs, Marsupials and Monotremes.

Briefly the mammals may be characterized as Vertebrates which suckle their young; this definition is enough to distinguish them from all other animals. In addition, mammals are usually characterized as being warm-blooded, or to be more accurate, by saying that their body temperature remains constant and independent of their surroundings. Although this is characteristic of the majority of mammals, it has been found that there are several important exceptions to this rule. In most mammals the body temperature ranges from 96° to 104°F, but certain mammals can tolerate considerably lower body temperatures; for example, the temperature of the spiny anteater may vary from 80° to 89°F; the temperature of bats decreases during ordinary sleep, and during hibernation their temperature decreases with that of the surroundings and may even drop below freezing-point.

If one wishes to be more specific about the characters distinguishing mammals from the other Vertebrates, one may say that they are typically land animals; their skin is more or less closely covered with hairs; they breathe with lungs, even when they spend their lives in the water; the heart has four chambers and the chest cavity is completely divided from the abdominal cavity by a partition known as the diaphragm. Mammals derive their name from the fact that the young as infants feed on the milk produced by the milk glands or mammae of the mother.

All the mammals illustrated in this book have their English name printed in bold type, followed by the scientific name in italics. Those mammals which are not illustrated, but which it has been thought desirable to mention, have their names printed in italics without the scientific name. In most cases the name of an animal is followed by two or three figures; the first gives the length of the animal from the tip of the snout to the root of the tail, the second gives the length of the tail, whilst the third figure gives the shoulder height. For example, **Lion**, *Felis leo*, $78 + 37 \perp 41$ in., gives successively the English name, the scientific name, the length from snout to tail base, the length of the tail, and the shoulder height. The scientific names used are those which are now generally accepted by zoologists, but here again there are still differences of opinion in regard to the correct names for some of the species.

The illustrations were completed before the preparation of the text and so the selection of species was decided beforehand. Some would, perhaps, have preferred more deer and fewer monkeys; to others there may seem to be a large number of rodents and rather few whales. In fact there are scarcely two persons who would make the same selection and the artist Wilhelm Eigener has on the whole included those illustrations which one might reasonably expect within the compass of a book of this size.

The following books may be of interest to those who wish to carry further their studies on mammals.

F. Boulière (1955) *The Natural History of Mammals.* London. Harrap.
L. Harrison Matthews (1952) *British Mammals.* London. Collins.
V. B. Scheffer (1958) *Seals, Sea Lions and Walruses.* London. Oxford University Press.
J. Z. Young (1957) *The Life of Mammals.* Oxford. Clarendon Press.

Primates

THE apes, monkeys, tarsiers and lemuroids are all grouped together in the order Primates, and the most modern view is that the tree-shrews (here shown under the Insectivores, p. 189) should also be included in this group. Primates walk on the soles of the feet, their limbs are in most cases adapted for life in trees, and all five fingers and toes are usually well developed. In many the thumb and big toe–if present–are separated from and more freely movable than the remaining fingers and toes (they are then said to be opposable); in this way the hands and feet are well adapted as gripping organs. The female usually bears only one young at a time.

Monkeys

The term *Monkeys* is here used in its widest sense to include apes as well as the tailed monkeys. Man is descended not from living, but from extinct members of this group, so that he and the living monkeys have a common origin among long-extinct forms. Most monkeys have a dentition similar to that of man, with 2 front teeth, 1 canine and 5 cheek teeth (premolars and molars) in each half of each jaw, making a total of 32 teeth; in most of the males the canines are longer than the other teeth and thus resemble those of the Carnivores. Many monkeys have jaw pouches to keep food in, and throat sacs which can strengthen the voice, but it is only the capuchins and gibbons which can produce a kind of song. The muscles of the face are more highly developed than in other mammals and allow the changes which we call facial expression. The eyes, which are their most important sense organs, are close together and face forwards so that the animals can see objects in perspective. In the skull as a whole the snout part is less prominent and the braincase more so than in other mammals. The brain is not as large, in absolute size, as it is in man. For example, a large gorilla has a smaller brain than a human infant; on the other hand there are certain small South American monkeys with relatively large heads, which have a proportionately larger brain than man. With the help of various sounds the apes and monkeys are able to express their mood, but they have no speech centre in the brain and are incapable of speech in the human sense. Apart from the elephants, monkeys and apes are the only mammals which can use tools. For instance gorillas

9

break off branches and twigs which they use to chase flies away, and baboons use stones as weapons against their enemies. It is rare for monkeys to live alone or in pairs; as a rule they go about in families or in large herds with the strongest male acting as leader, and he claims and receives absolute obedience from all the others, of both sexes. Polygamy is general throughout the group. In most species the period of gestation is about 7 months, and breeding does not appear to be restricted to any particular time of year. Monkeys can reach a good age; in zoos it is not unusual for occasional specimens to be over 20 years, and there are records of a mandrill of over 40 and a baboon of 45. One sometimes sees monkeys sitting and searching among their own fur or that of their companions; they are often said to be searching for fleas, but in fact they seldom have external parasites and in most cases they are hunting for the dry and salty pieces of skin which flake off. Monkeys are almost exclusively tropical forest animals, although some have become adapted for living on cliffs and have extended their range northwards. They feed principally on fruits and other plant material, but may also eat insects, eggs and birds.

Old World Monkeys

The monkeys (in the wide sense) may be subdivided into the narrow-nosed Old World monkeys, which include the apes, and the broad-nosed New World monkeys.

In the Old World monkeys the nostrils lie close to each other. The tail varies in length and may be lacking, but is never prehensile. They live almost exclusively in Africa, southern Asia and the East Indies. They can be divided into four families–man, the anthropoid apes, the gibbons and the cercopitheques (the latter including langurs, colobus, baboonlike monkeys and guenons).

Anthropoid Apes

The anthropoids show many points of resemblance to man; amongst these are the broad barrel-shaped chest, which does not occur amongst other monkeys, the lack of an external tail, and the presence of a vermiform appendix. Unlike the other monkeys they have neither cheek pouches nor bare patches on the buttocks. Anthropoids live in the trees and as an adaptation to this the arms–especially the forearms–are very long, longer than the legs, and well adapted for climbing. When they come down to the ground, they do not move about like other mammals, but walk on the feet and on the knuckles of the hands; sometimes they walk upright for short distances. The neck is so short that the head almost sits in between the shoulders; in the young the skull is smooth and round as in man, but later it grows ridges over the eyebrows, on the crown and back of the head, and on the jaws, but there is no resulting increase in the capacity of the cranium or brain-case. Quite rightly the anthropoids are regarded as the most intelligent of the animals; their intelligence is however scarcely greater than that of a small child, but it varies considerably between individuals. As a rule the face is naked, but the part without hairs is smaller than in man. The anthropoids are the

Gorilla, male, female and young

largest of the primates. The gorilla and chimpanzee live in the central and western part of central Africa and the orangutan in Sumatra and Borneo.

Gorilla, *Gorilla gorilla,* 78-90 in. (the anthropoids are measured from the top of the head to the heel and not, like most of the other mammals, from the tip of the snout to the root of the tail). The male, which is much more powerful than the female, may have a shoulder breadth of over 39 in., an arm span of nearly 120 in. and a weight of 550-700 lb. The gorilla lives in the tropical forests of central Africa, particularly in French equatorial Africa and the Congo; it does not however, have an equal distribution within this region, but occurs in limited areas in the forest–the so-called gorilla islands. The rigde on the top and back of the head is very strongly developed, especially in the old males. The ridges over the eyes and the nostrils are large in both sexes; the nose is flat, and in the male there is a broad channel running down its ridge. The central part of the face is naked, and the breast and belly have a sparse covering of hairs; elsewhere the coat is thick, coarse and black, with a rust-red tinge. Young gorillas are matt-black, but old ones eventually become grey-haired. The relatively short arms and the broad hands with short fingers are not particularly well adapted for climbing, and gorillas remain on the ground more than other anthropoids. When standing or walking they rest on the knuckles of the hands. Gorillas wander about either in family parties or in groups of up to 20-30 animals. The young remain with the family until they become sexually mature at about 14 years of age or perhaps earlier. In the evening they build a sleeping-place of branches and twigs up in the trees or in bushes. In most cases such a nest is only used for one night. The female usually sleeps in a nest on the ground or sits with her back against the tree where the family in spending the night. The food consists of fruits, including bananas, together with sweet-corn and sugar-cane; in addition the young eat insects, young birds and eggs. When excited, a gorilla stands up on the hind legs, bares its teeth, growls and drums on its breast with the fists.

The *Mountain Gorilla,* 60-70 in., should possibly be regarded as only a special race of gorilla. It has shorter arms and is completely black, except that the old males have pale grey backs. The coat is thicker and longer, as might be expected in an animal which lives at high altitudes in places where there may be night frosts. Mountain gorillas are found near Lake Tanganyika and Lake Albert.

Chimpanzee, *Anthropopithecus troglodytes (=Pan satyrus),* 47-65 in.; the male is more powerful than the female and weighs 110-165 lb. Chimpanzees live in the tropical forests of the western part of central Africa, and are much commoner than gorillas–they sometimes go about in groups of up to 50 animals–and are distributed over a much wider area. The arms are shorter than in the gorilla and the ears much larger, the hands much narrower, the fingers much longer and altogether the chimpanzee is a much better climber. They usually walk on all fours, but they can walk upright, often with the arms held above the head for the sake of balance. In the new-born chimpanzee the naked face, hands and feet are flesh-coloured. As they grow older the skin becomes darker and ends by being black. At an age of ten to twelve, when they are sexually mature, they are usually completely black; old chimpanzees may become grey-haired. Like the gorilla, chimpanzees feed on fruits and build nests in the trees. The gestation period is about 8½ months and the new-born chimp weighs 4-5 lb. There are at least four races or subspecies of chimpanzee.

The *Gambia Chimpanzee,* from Guinea Sierra Leone and Liberia, has a greyish-white beard when adult, a hairy forehead and a parting down the middle of the scalp.

The *Tschego Chim-panzee* lives further to the east, from the Niger and the Gulf of Guinea to the Ubangi and Con-go Rivers. It has almost no growth of beard, a smooth forehead, and retains its pale face col-our until it is three to four years old.

The *Long-haired Chim-panzee* lives from the River Ubangi to Lakes Albert and Victoria; the male has a well-devel-oped beard all round the face.

The *Pigmy Chimpan-zee,* first found in 1928, lives south of the Congo River. It becomes dark-faced at an early age.

Orang-utan, *Pongo pyg-maeus,* 47-66 in., 150-220 lb.; the male is consider-ably larger than the fe-male, and its arm span may be over 100 in. The orang is found in the tropical forest areas of Borneo and nothern Su-matra, usually living alone or in family par-ties; the young remain with the family until they are ten years old, by which time they have become sexually mature. The coarse, red-brown coat is thin on the breast, but long and thick on the shoulders, arms and legs and along the sides of the body; exception-ally the hair may be 18

Chimpanzee. (above) *male;* (below) *female and young*

13

in. long. In the adults the hairs on the top of the head hang down over the forehead; old males usually have a full beard. The ears are very small —smaller in fact than in the other anthropoids. The arms and fingers are remarkably long, but the legs are proportionately short and the feet are long and narrow with in-turned soles. The old males develop enormous swollen folds on the cheeks which give the face a grotesquely flat appearance. Orangs spend almost the whole of their lives up in the trees. They live exclusively on fruits, including those which are sour and bitter, and sometimes visit native plantations to get hold of mangosteens and other fruits. They build nests in the trees like the other anthropoids. The period of gestation is nine months, and the new-born young weighs about 2½ lb.

Orang-utan. (above) *young and female;* (below) *male*

Gibbons. (l-r) *Lar Gibbon, Hoolock Gibbon, pale form of Hoolock, Grey Gibbon*

Gibbons

The gibbons were once regarded as anthropoids, but they are now classified as a separate family, with two genera. They are all rather similar to each other, except in coloration, which may vary within a single species and even between individuals. They are about 20-40 in. tall and have unusually long arms and fingers, so that the arm span may be over 75 in.; yet they weigh only 13-25 lb. The face is naked, and the small head is rounded and lacks bony ridges. The back legs are short and there is no external tail, but in the adults there are small bare patches on the buttocks. The period of gestation is 7-8 months and the young become sexually mature when they are 5-8 years old. Gibbons live in pairs or in flocks, in the tropical forests of Indonesia and south-east Asia, and feed on fruits and fresh shoots. They swing by the arms from branch to branch–at great speed–and in this way may sometimes make three or four jumps one after the other, each time covering up to 30 ft. On the ground they walk, or rather totter, with the arms held out at the sides or partly above the head as a means of balance. In the morning and evening their loud song reverberates through the forest. They drink by dipping the hand into water and supping it up, or by holding it above the head so that the water drips from the fingers into the mouth.

Lar Gibbon, *Hylobates lar.* The face is surrounded by greyish-white hairs; the hands and feet are white and the rest of the body is greyish-black or brown. They live in Burma, Siam, Malaya, Sumatra, Java and Borneo.

Hoolock Gibbon, *Hylobates hoolock.* The male has a white band on the forehead, but is otherwise black; in the female the coat is brownish or greyish-yellow. They live in the north-western parts of south-east Asia.

Grey Gibbon, *Hylobates cinereus.* In this species the body is grey with a lighter edging round the pale face. They live in Java and Borneo.

The other gibbon genus, *Symphalangus,* contains the *Siamang,* which has a shiny black coat, and a very large throat sac; the second and third toes are united. Siamangs are found in the mountain forests of Malaya and Sumatra.

Cercopitheques

The cercopitheques form the third and last family of Old World monkeys. This family is very rich in species, which differ from those in the last two families in that the chest is compressed and the front limbs are not longer than the hind limbs; the tail is usually long. They walk on all fours, usually on the palms of the hands and the soles of the feet. There are two sub-families, one containing the langurs, colobus monkeys, leaf monkeys and the proboscis monkey, whilst the other has the baboons, mandrills, mangabeys, macaques and guenons.

The monkeys in the first sub-family have a round skull, short jaws and feed principally on leaves. They are mostly

Langur or Hanuman Monkey

slower-moving than the other monkeys and may often sit still, quietly digesting, for hours at a time.

The *Langurs* are slender monkeys with a small head, short muzzle and no cheek pouches.

Langur or **Hanuman Monkey,** *Semnopithecus entellus,* 25+40 in. One of the commonest monkeys in India and regarded as sacred by the Hindus. In many places it has become an agricultural pest.

Proboscis Monkey, Nasalis larvatus, 27+31 in. The old males have large swollen noses which hang down over the mouth. Found in the tropical forests of Borneo, often along the rivers.

The *Colobus Monkeys* have long-haired coats and short fore-limbs.

Guereza Monkey, *Colobus abyssinicus,*
29+39 in. Extremely agile monkeys
which live in the tropical forests of Africa, often at altitudes of 6,000 to 9,000 ft.
above sea level. The Africans have used
the beautifully marked skin for shields
and ankle decorations, and, as the pelts
have also been exported to Europe,
many of the races of colobus have become very rare. In West Africa there is
a completely black form of colobus monkey, known as the *Black Guereza,* which
has long erect hairs on the forehead.

Guereza Monkey

Proboscis Monkey. (above) *female and
young;* (below) *male*

In the second sub-family of cercopitheques, we come first to the baboons
and mandrills, which live in Arabia and
Africa. In these the limbs are not particularly long, and the difference between
front and back legs is about the same as
in the carnivores and other typical mammals. The hands are powerful, with short
fingers and a well-developed thumb, the
body is short and stocky, and the head
is large with a small brain-case, strongly
developed ridges over the eyes, an elongated muzzle and jaws, and large cheek
pouches. At night most of them go up
into cliffs or deep into caves to get shelter from their main enemy, the leopard.
The old males, which rule over a whole
herd, have strong, pointed canine teeth
and can hold their own with a leopard.
Such a male will fight off other males

17

Mandrill, young, female and male

as long as his strength lasts. The females are smaller and weaker than the males, and there is nearly always polygamy.

The *Mandrills* have a strikingly large head with powerful jaws. The nostrils are large and have a thickening of cartilage along the edge, and the naked cheeks are swollen and longitudinally ridged. They have short tails, and in general the differences between the sexes are greater than in other monkeys. They live in the tropical forests of Africa, between the Niger and Congo, in the region of the world where the greatest number of monkeys is found; but we do not know exactly how far they extend to the east.

Mandrill, *Mandrillus sphinx,* $37 + 1\frac{1}{2}$ in. The adult male has the most brightly coloured face of all mammals; the swollen cheeks are sky-blue, the nose is bright red, the moustache white, and the pointed beard, which continues into a thick collar almost the whole way round the neck, is reddish-yellow. The coat of the male is dark brown with a greenish sheen on the back, and the belly is greyish. The naked rump is red near the stumpy tail, grading into red-violet and pale blue. Mandrills live on the floor of the forest during the day, but climb up into the trees at night; they feed on grass, roots and nuts, but may also take insects and other small animals.

The *Drill* is smaller than the mandrill and has a black face with a white beard on the cheeks. The bare patches on the buttocks are red, and the coat is black on the back and white on the belly.

Sacred Baboon, *Papio hamadryas,* $27 + 11$ in. The striking shoulder cape of the male is formed by the long hairs on the front part of the body. The buttock patches are bright red. Sacred baboons live principally in Somaliland and the coastal regions on both sides of the Red Sea, usually keeping up in the mountains at altitudes of 3,000-6,000 ft., but sometimes going right down to the sea. They go about in herds of 150-300 animals, of which 10-30 are fully-grown males, about twice as many are females, and the remainder are half-grown and young. They spend the night in holes in cliffs, search for food during the forenoon, and move off to drinking-places as night approaches. They are often seen turning over stones in search of insects, larvae and other small animals, and will also eat birds' eggs and chicks, but they live mainly on fruits, roots and grass. Whilst a herd is feeding they always have lookouts posted on the surrounding rocks. The gestation period is about 7 months. They are adult when 8-9 years, and may exceptionally live for 24 years or more.

The Yellow Baboon, *Papio richei,* 29½ + 13 in., which is strikingly long in the limbs, lives on the steppes in Somaliland. The colour is more or less yellowish with the face darkish.

The *Mangabeys* live in herds in the tropical forests of Africa south of the Sahara. They are long-legged and usually dark in colour, and have large bare patches on the buttocks. The *Sooty Mangabey* from West Africa has a naked, yellowish-brown face and white eyelids. The *Black Mangabey* from the Upper Congo has black fur with a tuft of hairs on the top of the head.

The *Macaques* form another large group of monkeys, having the following characters in common: the muzzle is long, but not so long and powerful as in the baboons; the nostrils are not at the tip of

Baboons.
(left, above) *Sacred Baboon, male,* (below) *female;* (right) *Yellow Baboon, female and male*

the snout but a little above it; and the thumb and big toe are developed as in other tree-climbing monkeys. They feed on roots, buds, leaves, fruits and seeds, as well as on insects, worms, larvae, eggs, chicks and other small animals. With the exception of the Barbary ape the macaques all live in southern and eastern Asia.

Stump-tailed Macaque, *Macaca arctoides,* 19 in. tall. The tail is very short and naked, the coat is brown and the area round the eyes is red with black markings. Brown macaques live in Burma, Siam, Indo-China and South China, keeping up in the mountains.

Black Ape, *Cynopithecus niger,* 24 in. tall. These have prominent ridges over the eyes, a conspicuous tuft of hair on the top of the head and a very short stumpy tail. They live in herds in the island of Celebes. In spite of their English name neither they, nor the Barbary ape, are, of course, true apes.

Moor Monkey, *Macaca maurus,* 19 in. tall. The tail is very short and stumpy, and the coat is chocolate-brown in the young, but later becomes black, often with grey limbs. Moor monkeys live in Celebes.

Barbary Ape, *Macaca sylvana,* 19 in. tall. The tail exists only as a vestige. This is the only monkey in Europe, and the only one which lives in Africa north of the Sahara. In Europe they are restricted to the cliffs of Gibraltar, where they were almost exterminated during the Second World War. They were saved by Winston Churchill giving orders for fresh stock to be brought in from northwest Africa, where they are found in the mountains of Morocco, and in Algeria. At one time it was thought that Barbary apes had been originally introduced into Europe, but this view seems to be mistaken, as fossil remains of similar monkeys have been found in France, Switzerland and Italy; also, of course, Spain and Morocco were at one time joined to each other. The coat is reddish-brown, with the feet and hands dirty brown. The Barbary ape feeds on plants and

Macaques. (l-r) Stump-tailed Macaque, Black Ape, Moor Monkey

Barbary Ape

insects and turns over stones in search of scorpions and other small animals, and is evidently a hardier animal than the other macaques.

The *Pigtailed Monkey* has a tail 5-6 in. long, and is somewhat like a baboon. It lives in south-east Asia, Sumatra and Borneo. The Malays train it to pluck and knock down coconuts. It is said to be able to distinguish between ripe and unripe coconuts.

The *Lion-tailed Monkey* from India is black with a prominent circle of grey hair round the face; the tail is fairly long and has a tuft at the end.

Bonnet Monkey, *Macaca radiata,* 19+ 23 in. The forehead and cheeks are almost naked, and the face and ears, together with the naked parts of the hands and feet, are pale flesh-coloured. The

hairs on the top of the head grow out from a central naked parting and produce a hat-like effect. Bonnet monkeys live in southern India, even in towns, and the natives regard them as sacred animals. In fact they play the same role in southern India that the Hanuman monkey does further to the north. The closely-related *Toque Monkey* of Ceylon also has a crown of hair radiating from the top of the head.

Rhesus Monkey, *Macaca mulatta,* 24+ 10 in. The grey-green back grades into reddish-yellow on the hindquarters, the belly is white and the tail grey-green. The face, ears and hands are pale copper-coloured. When frightened a rhesus monkey becomes red in the face. They live on the ground, in trees and on cliffs, and near water they can often be seen ducking and swimming. Rhesus monkeys usually live in large herds and are just as troublesome and importunate as bonnet monkeys. Rudyard Kipling has written of them as always fighting, stealing and teasing each other. There are several records of them having thrown stones down on to men–a first attempt at using tools. Rhesus monkeys live in the central and northern parts of India and east to Siam and Indo-China; in the Himalayas they may occur up to heights of 6,000-9,000 ft. They are common in zoos, and once established may breed frequently and live long, sometimes for twenty to thirty years.

Kra or **Crab-eating Monkeys,** *Macaca irus,* 20+22 in. A very common monkey in south-east Asia and the East Indies. They live in small herds in mountain areas, in jungle and along rivers, and also come down to the coast to eat crabs and bivalve molluscs; otherwise they feed mainly on fruit and other plant food.

Patas Monkey, *Erythrocebus patas,* 21 + 23 in. There are several races of Patas monkey in tropical Africa. Unlike the guenons (see below) they are not arboreal, and spend most of the time on the ground. The snout or muzzle is black in the young, but may become completely white in the adults. Patas monkeys form a kind of transition between the macaques and the guenons.

Bonnet Monkey and Rhesus Monkey

Kra or Crab-eating Monkey (above);
Patas Monkey (below)

The *Guenons* form a large group of monkeys, all of which are rather similar to each other. On the average they are about 21 in. long, with a tail 29 in. long, and they are good climbers and jumpers. They have a round head with a projecting muzzle and ridges over the eyes; and they have large cheek pouches. The hands are small and slender, with proportionately long thumbs, and are thus well adapted for plucking fruits and buds and for getting eggs and insects out of holes. Guenons live gregariously in the tree-tops in tropical Africa, preferably in the neighbourhood of rivers or other open water. They betray themselves by their constant tumult and screaming. Sometimes they come down to the ground to raid cultivated land, with an old and experienced male leading the foray. At first the pack approaches with great care, leaping from tree to tree. The old male keeps well in the van, with the others following him in a long string and choosing exactly the same route among the branches. Now and then the leader climbs up to the top of a tree to prospect, and if the coast is clear he gives a sign for them all to move on. Eventually they climb down from the trees and move out into the open fields. First they tear off the maize cobs, pick off the grains and fill their jaw pouches until they are bulging; later on they become more fastidious and take only one or two seeds from each cob and throw the rest away. Thus it is quite understandable that this wanton and wasteful destruction makes them hated by the natives. However, apart from man, guenon monkeys have few enemies, for they are very fast both on the ground and in the trees; exceptionally one may be caught by a leopard, and sometimes a python manages to get one. Guenons do not appear to have any special breeding season, for one can see young of almost all ages in the same herd.

Green Monkey, *Cercopithecus sabaeus,* from Sierra Leone, Liberia and Senegal. The back is greenish, the belly whitish, the limbs grey and the face black. Green monkeys have been introduced into Barbados and St. Kitts in the West Indies. The closely related *Grivet Monkey* lives in Abyssinia and Sudan.

Red-eared Monkey, *Cercopithecus cephus erythrotis.* The ear tufts are reddish-brown, the nose pink, the hands and feet black and the beard yellowish. They live in the Cameroons and on Fernando Po. The closely-related *Moustached Monkey* has white on the upper lip, a blue face and yellow whiskers.

Guenons. (above) *Red-eared Monkey;* (centre) *Green Monkey;* (below) *Wolf's Monkey*

Mona Monkey, *Cercopithecus mona.* A reddish monkey with a white throat, red round the mouth and a red tuft at the tips of the ears. It lives in West Africa. The specimen illustrated is **Wolf's Monkey,** a subspecies of *Cercopithecus mona,* from the Congo.

Diana Monkey, *Cercopithecus diana,* is black with red-brown along the back

Guenons. (top, right) *Diana Monkey,* (left) *Diadem Monkey;* (bottom, left) *Sykes' Monkey,* (right) *Brazza's Monkey*

and white on the neck, breast and upper sides of the arms, and a white stripe on the hips. It lives in the forests of West Africa.

Diadem Monkey, *Cercopithecus leucampyx leucampyx.* This is one of several subspecies of a very variable species. The white band on the forehead is usually narrower than shown in the

illustration. It lives in tropical West Africa.

Sykes' Monkey, *Cercopithecus leucampyx albogularis.* Closely related to the diadem monkey, but somewhat larger. It is distributed from the east coast of Africa to the Congo; there are several races.

Brazza's Monkey, *Cercopithecus brazzae.* The forehead is strikingly marked with a broad reddish-yellow band, behind which there is a black band running right over the top of the head. The chin and lower part of the cheeks are white. They live in French equatorial Africa and the Congo.

New World Monkeys

The New World monkeys form the second of the main groups of monkeys; the nostrils face outwards and are separated by a broad bridge of cartilage. The tail is well developed and sometimes prehensile, and they have no bare patches on the buttocks and no jaw pouches. The thumb is weakly developed or may be lacking. The New World monkeys are found exclusively in the rain forests of South and Central America, where they live up in the tree-tops. There are two families: the Cebidae, which contains the spider-monkeys, capuchins, howling monkeys and douroucoulis, and the Hapalidae, with the marmosets and tamarin monkeys.

Cebidae

The *Spider Monkeys* have the most specialized prehensile tail of all mammals. Apart from this their very long limbs—of which the arms are the longest—remind one of the gibbons. On the underside of the long tail there is a naked part towards the end, which looks rather like the sole of the foot; this area is very sensitive and functions as a kind of fifth limb. Spider monkeys can walk on the hind legs but they very rarely come down to the ground. Even when drinking they will only climb out on a low overhanging branch, hang down by the tail and scoop up the water with one of the hands. The thumb is entirely lacking, but the other fingers are remarkably long and form a kind of long hook rather than a true gripping organ. Spider monkeys occur from Mexico to Uruguay, often in small herds of 6-12 animals. In spite of their long prehensile tails they are not such outstanding acrobats as the best of the Old World monkeys, although one can sometimes see a whole herd of them suspended by their tails. The chin and throat are naked and it is characteristic that the hair on the top of the head lies forwards. There are several species.

Geoffroy's Spider Monkey, *Ateles geoffroyi,* 25+25 in. The coat may be either yellowish-grey or reddish-brown. They live in Central America and the northern part of South America.

The *Woolly Monkeys* form a separate genus; they have thumbs and rather short arms, and a prehensile tail.

Humboldt's Woolly Monkey, *Lagothrix humboldti,* 27+27 in. This is one of the largest of the New World monkeys. Humboldt's woolly monkeys are gregarious animals which feed almost entirely on fruits. They are found in Colombia and the northern parts of Peru and Brazil.

Geoffroy's Spider Monkey

The *Capuchin Monkeys* are so named because the thick erect hair on the head has the appearance of a monk's cowl. The tail is long and hairy, but not truly prehensile, although it can be curled in a spiral around the branches of trees. The fur may be coarse or woolly, according to the species. Capuchins have

Humboldt's Woolly Monkey

a soft voice and they spend almost their whole lives up in the trees where they feed mainly on fruits, but they come down for water.

White-throated Capuchin Monkey, *Cebus capucina,* 17½ + 25 in. This species lives in Nicaragua, Costa Rica, Panama and Colombia. It has a flesh-coloured face, white cheeks, throat, chest and shoulders, but is otherwise black.

Brown Capuchin Monkey, *Cebus fatuellus,* 17 + 16 in. The fur is brownish, with the limbs and tail rather darker. It lives in Guiana, Brazil and Colombia.

Capuchin Monkeys. (top, left) *White-throated Capuchin Monkey,* (right) *Squirrel Monkey;* (below) *Brown Capuchin Monkey*

The *Squirrel Monkeys* are only about the size of a squirrel, and the tail is not really prehensile.

Squirrel Monkey, *Saimiri sciurea,* 11 + 19 in. The coat is yellowish with a tinge of green. The eyes in the little round head are rather large, and the area round the nose and mouth is black. They live in the forests of Guiana, often in groups of over a hundred, and are also found in Venezuela and Colombia.

The *Howling Monkeys,* 25 + 27 in., have a tail which is naked on the underside towards the tip. The very large throat sacs and a modification of the hyoid bone together form a powerful resonating organ, so that their voice can be heard for miles in the forest. The coat is thick and the beard long. Howling monkeys go about in small herds of 5-10 animals with an old male as leader, and feed almost exclusively on leaves. They climb with a slow creeping movement from branch to branch, using the prehensile tail as an emergency anchor. Their most dangerous enemies are birds of prey.

Black Howling Monkey, *Alouatta caraya.* Distributed from Paraguay to Ecuador. The adult males are coal-black, the females and the young brownish or yellowish.

Red Howling Monkey, *Alouatta seniculus,* lives in Ecuador, western Brazil, Venezuela and Colombia, and is coppery-red in colour.

The *Douroucoulis* are the only monkeys which are nocturnal. The eyes are very large and may shine brighter than a

Black Howling Monkey, male and female
(above); *Red Howling Monkey* (below)

Marmosets

The ears are conspicuous and usually have long hairy tufts. The fingers and toes have claw-like nails, except for the thumbs which are opposable and have ordinary nails. The tail is long and bushy, but is never prehensile. They can sit up on their hind legs like a squirrel and hold their food between the front paws.

The marmosets are essentially forest animals. They feed mainly on insects but also eat fruits, leaves and flowers. Their sharp teeth are rather like those of the Insectivores. They spend the day up in the tree-tops; by night they gather together in big clusters in hollow trees. As with other monkeys, the female usually has one young at a time, but quite often two or three, and in such cases she usually rears only one of them. Their worst enemies are birds of prey.

Common Marmoset, *Hapale jacchus,* 9 + 12 in. The fur is long and soft, and coloured yellowish-grey on the back

cat's eyes in the darkness. The coat is soft and the tail is bushy and not prehensile.

Three-banded Douroucouli or **Night Ape,** *Aotus trivirgatus,* 13 + 19 in. The grey-brown coat is rusty on the belly and the tip of the tail is black. During the day they sleep in holes in trees, but move around at night looking for insects, sleeping birds and fruit. They live in Brazil, Guiana and Peru.

Three-banded Douroucouli or Night Ape

with light and dark cross-bands, which continue right back to the tip of the tail. They are native to an island in the mouth of the Amazon.

The *Black-eared Marmoset* of eastern Brazil is probably just a race of the common marmoset. It has a darker coat and dark-brown ear tufts.

Black-tailed Marmoset, *Hapale argentata,* 8 + 10 in. The coat is silvery-white, the tail matt-black and the face flesh-coloured. This is a rare animal which lives in the forests of Brazil and Bolivia.

The next two species differ from the other marmosets in having long canine teeth and lacking the ear tufts.

Pinché Marmoset (left); *Silky Marmoset* (right)

Common Marmoset (left); *Black-tailed Marmoset* (right)

Pinché Marmoset, *Oedipomidas oedipus,* 12 + 15 in. The coat is brown, with a long cape hanging down from the back of the head. No other mammal has a voice which so resembles the song of a bird; it can produce long, pure piping notes and rapid trills. It lives in eastern Colombia.

Silky Marmoset, *Leontocebus rosalia,* 12 + 17 in. The fur is chiefly reddish-yellow, and very soft and luxuriant. Silky marmosets live in forest country, and feed on fruits, insects and other small animals, which they find up in the tree-tops. They are distributed from Brazil to Colombia and Panama.

Tarsier

Tarsiers

Tarsier, *Tarsius tarsius,* 7+9 in. A distinctive and characteristic animal that is classified in its own sub-order, of equal importance with the monkeys and lemurs. The tarsier has some structural characters in common with these two groups, but differs from them in the gigantic eyes and the much lengthened base of the foot. The second and third toes on the hind feet have upward-facing claws, but all the other digits have nails, and all of them have small flat pads at the tips. The coat is woolly, the tail is very long and ends in a tuft, and the molar teeth have numerous points. When moving about in the trees a tarsier looks like a rat-sized tree-frog. The likeness is enhanced by the fact that the pads on the toes function as adhesion discs. Tarsiers are nocturnal animals and sleep by day in hollow trees. They live mainly on insects and can eat 30-40 grasshoppers in a single meal. There are various races in Borneo, Sumatra and the Philippines.

Lemuroids

The animals in this group resemble monkeys in the development of the hands and feet for climbing, but differ from them, amongst other things, in having a long-snouted, dog-like head. All the fingers have flat nails, except the second toe on each foot which has a claw. The front legs are shorter than the back legs, the thumb is well developed and the tail is usually long. As a rule there is a space between the two middle incisor teeth in the upper jaw, and the narrow front teeth of the lower jaw are turned forwards; at one time it was thought that these teeth were adapted for combing the fur, but in fact they are probably used more for rasping at the fruit which the animals eat. Apart from fruit, they live mainly on insects. They usually bear one young at a time, although sometimes there may be two or three. The new-born young hooks itself fast to the mother, and is thus dragged around clinging to her belly. There are two nipples on the breast and often one or two pairs on the belly. They are small to medium-sized, woolly-haired, nocturnal animals with large, close-set eyes. They live gregariously in the tree-tops; some of them can purr like a cat. Many of them get through periods of bad drought by a kind of aestivation. Their area of distribution is first and foremost Madagascar, but there are also species in Africa south of the Sahara, in India and Ceylon, south-east Asia, the East Indies and the Philippines. They play an important role in Madagascar, where about half the island's mammal fauna consists of lemuroids.

Bush-tailed or Great Galago

Lorisids

The species in this family have short or rudimentary index fingers and they all come from Asia or Africa, but are absent from Madagascar. There are two sub-families, the galagos and the lorises.

The *Galagos*, or *Bush Babies*, have large naked, thin-skinned ears, which have muscles, so that the animals can fold them in and close the ear-hole during the day while they are sleeping. The fur is thick and soft, the tail long and bushy, and the hind legs are long and adapted for jumping.

Bush-tailed or **Great Galago,** *Galago crassicaudatus,* 13+14 in. The coat is greyish yellow with a pale grey belly and ash-grey ears. By day they lie rolled up and fast asleep, but towards evening they wake up and start to climb around, sometimes with the head held downwards. They feed mainly on insects and small mammals. They live in Kenya,

Tanganyika, Nyasaland, Rhodesia and southwards to Natal.

The *Lorises* from India and the *Pottos* from Africa have short ears and reduced tails; they crawl slowly along the branches of trees and never jump. The thumb and big toe are well developed and opposable, so that the hands and feet can grip the branches like a pair of pincers.

Slender Loris, *Loris tardigradus,* 10 in. The soft fur varies in colour according to the race, but is most often reddish grey on the back and whitish yellow on the belly. They are able to approach their prey—often small birds—without making a sound; they also eat insects and fruits. They live in Ceylon and southern India.

Slender Loris (above), *and Slow Loris*

Slow Loris, *Nycticebus coucang,* 14 + ¼ in. The fur is mouse-grey in colour and there is a short tail; otherwise they are similar in appearance and habits to the slender loris. They are found from Bengal through Burma, Siam and Malaya to Java, Sumatra and Borneo.

Lemurs

The lemurs have well-developed index fingers and are found only in Madagascar. There are two sub-families, the true lemurs and the indris.

The *True Lemurs* have long fox-like snouts, big ear tufts and long tails. They go about mostly at night in the tree-tops, but may also be seen by day in small groups of 6-12 animals. They move so quickly from branch to branch that they almost look as though they were flying. Apart from small birds and insects they also eat dates and bananas, and sometimes come down to the ground to collect fallen fruits. They are hunted by the natives for their tasty flesh. As soon as the sun has set the lemurs start their harsh cries, which sound frightful, for a whole flock will scream at the same time. The period of gestation is 4-5 months and the young are born entirely naked. At first they hide away in the mother's thick fur, but after about six weeks they start to behave like the adults, although they remain in the care of the parents for some six months.

Ring-tailed Lemur, *Lemur catta,* 17 + 18 in. The body is greyish, the face, ears and belly are white and the tail has alternate black and white rings. They

Lemurids. (above) *Black Lemur, male and female;* (below) *Ring-tailed Lemur*

Lemurids. (above) *Ruffed Lemur;* (centre) *red form of Ruffed Lemur;* (below) *Indri*

31

live in the southern part of Madagascar, often in hilly districts with open woodland. Their voice is more cat-like than that of the other lemurs.

Black Lemur, *Lemur macaco,* 21 + 21½ in. Unlike the other lemuroids the sexes differ in colour, the male being coal-black and the female either brown or reddish. They have long ear tufts. Black lemurs live in the forests along the north-west coast of Madagascar and on the off-lying islands.

Ruffed Lemur, *Lemur variegatus,* 22 + 24 in. This is the largest of the lemurs and may be nearly as large as a fox. They live in the forests of east Madagascar. When sitting on a branch the tail may sometimes hang vertically down but at times is held up over the back like a squirrel's tail. The body is mottled black and white, with a red ruff and black feet and tail; there is also a race in which the body and ruff are red.

The *Indris* have a short tail and long legs with large hands and feet.

Indri, *Indri indri,* 31 + 1 in. This is the largest of all the lemuroids; the woolly fur is brownish-black with pale markings. Indris are found in the hilly country of eastern Madagascar; unlike most lemuroids they are diurnal, and feed mostly on fruits, but will also catch small birds.

Aye-aye

Aye-ayes

This family contains only a single species.

Aye-aye, *Daubentonia madagascariensis,* 17 + 21 in. The aye-aye is about the size of a cat and has large naked ears and a big bushy tail. There are only two large rodent-like teeth in each jaw, and there is a gap between these front teeth and the remainder of the dentition. When it has bitten a hole in a bamboo or sugarcane an aye-aye fetches out the pith with the long third finger of the hand. All the fingers and toes have claws except the thumbs and big toes. Aye-ayes live in bamboo thickets in eastern and north-western Madagascar and are rather rare. They are nocturnal and go about either alone or in pairs.

Ungulates

THE Ungulates form one of the largest groups of mammals, and they can be subdivided in a number of different ways. Some consider that they are a super-order containing five orders: the elephants, the sirenians (manatee and dugong), the hyraxes, the odd-toed ungulates and the even-toed ungulates. Others consider that only the odd-toed and the even-toed groups should be regarded as sub-orders of the Ungulates; this latter classification is adopted here.

Most of the Ungulates are large, fast-moving herbivores, well adapted for running fast over wide grassy plains when pursued by their enemies. The limbs are elongated, with the metacarpal bones of the fore-limbs and the metatarsals of the hind-limbs very long. The toes are sheathed with solid hoofs on which the animals walk, and the remainder of the foot does not touch the ground. The molar teeth usually have broad crowns and are well adapted for grinding and crushing the plant food so that the digestive juices can act upon as large a surface as possible. Ungulates are found wild, in all parts of the world, except in the Australasian region. They have supplied man with a large number of domestic animals.

Perissodactyls or Odd-toed Ungulates

The perissodactyls, or odd-toed ungulates, are characterized not so much by the number of the toes, but by the fact that the central axis of the foot runs through the third toe, which is much more powerful than the others; in the horses the third toe is the only one which touches the ground. Most of the animals in this group have three toes, that is, the 2nd, 3rd and 4th, the only exceptions being the tapirs which have four toes on the fore-limbs. They all have very mobile lips, and also two nipples situated between the hind-limbs. The sub-order consists of three families, the rhinoceroses, the tapirs and the horses. At first sight it might seem that these three families have little in common, but in fact there are many extinct species which provide transitional forms between them and show that the classification is in fact justified.

Rhinoceroses

The large stout body with thick powerful legs, the thick, rigid and almost naked skin and the horn or horns on the snout make it easy to distinguish the rhinoceroses from all other mammals.

Great Indian Rhinoceros

The mouth is relatively small and the eyes very small, so that rhinoceroses have poor sight. On the other hand, their hearing is good; the ear is shaped like a cornet, with hairs along the edges. The tail has a little tuft of hairs at the tip. Each limb has three toes with hooves, but without pads. The horns are made up of fine horny threads which form a compact mass. The incisor and canine teeth are more or less reduced. Rhinoceroses usually live together in pairs with their young. The period of gestation is 17-18 months and the young rhino is fed by the mother until it is almost two years old. There are five species of living rhinoceros in Africa and south-east Asia.

Great Indian Rhinoceros, *Rhinoceros unicornis,* 120+23 ⊥ 67 in. The skin, which is even thicker than that of an elephant, is much folded around and on top of the neck and on each shoulder; on the hind-quarters and on each thigh there are large shields of skin separated by deep folds. The horn is longer and more pointed in the female than in the male and may reach a length of 15 in. They feed mainly on grass. The Indian rhinoceros is now found only in a narrow belt of country along the foot of the Himalayas from Nepal to Assam. The total surviving population has recently been reckoned at about 250 animals.

An Indian rhinoceros weighs about two tons.

The *Javan Rhinoceros* is somewhat smaller. It lives in Malaya and Java and there are only about 30-40 of them left in protected areas.

The *Sumatran Rhinoceros* is the smallest species, and it has two horns. It is native to Burma, Malaya, Sumatra and Borneo, but was almost completely exterminated during the last war, and its present status is still in some doubt.

Black Rhinoceros, *Diceros bicornis,* 95 + 23 ⊥ 63 in. The front horn is the longest and may measure 28-30 in., and exceptionally even longer. The upper lip is drawn out to a point, and there are no incisor or canine teeth. Black rhinos browse on leaves and shoots in the savannah country of Africa south of the Sahara, but have been much hunted both by Europeans and Africans.

The *White Rhinoceros* is larger and its broad upper lip, without a point, is well adapted for grazing. It has two horns and lacks both incisor and canine teeth. This is a rare animal found principally in southern Sudan and in a nature reserve in Natal. Both the black and the white rhinoceros are grey in colour. Some consider that the Boers used the name white rhinoceros because the first specimens seen had been wallowing in mud which had dried on them and given them a whitish appearance. It is more likely that the name arose through a confusion of the world *wide,* from the form of the upper lip, with *white.*

Black Rhinoceros

South American Tapir

Tapirs

Tapirs may be recognized by the short movable trunk. The neck is short and the skin is thick and has short hairs. The toes are more spread than in the rhinoceros and there are four toes on the front feet and three on the hind feet. Tapirs are found in Central and South America and in south-east Asia, and are usually active in twilight or at night. They live a sheltered forest life and travel by preference along well-known routes. At the approach of danger they flee towards water, for they are good swimmers and divers. Hearing and smell are well developed, but their eyesight is poor. A family of tapirs usually keeps together, although the old males often wander off alone. The period of gestation is 13 months, and the young are always dark with light spots and stripes. The females are often more powerful than the males.

Malayan Tapir

South American Tapir, *Tapirus terrestris*, $74+4 \perp 39$ in. The body is brownish-black in colour and there is a short stiff mane along the back of the neck. The young animal is dark with white spots on the top of the head and four rows of pale longitudinal stripes or spots along the side of the body. The stripes and spots become less obvious as the animal grows and after two years they have completely disappeared. Tapirs are found in South America, from Venezuela and the Guianas through Brazil to Paraguay and northern Argentina. They are animals of the dusk and live in those parts of the forest which are near to water. They feed on leaves, including palm leaves, and may also do damage to plantations by eating sugar-cane and fruits.

The *Mountain Tapir*, of Colombia, Ecuador and Peru, lives in the Andes at heights of 6,000–9,000 ft., and has a thick greyish-black coat. In addition there are two rare species of tapir in Central America.

Malayan Tapir, Tapirus indicus, $95+3 \perp 36$ in. The body is brownish-black in colour with a greyish-white area covering most of the back. The young animal is black with rows of longitudinal white stripes and spots, which disappear in the course of about ten months. They are nocturnal and live in southern Burma and Siam, Malaya and Sumatra, where they are still common in many places.

Horses

Horses form the largest of the three families of odd-toed ungulates and comprise the zebras, the asses, the Asiatic asses and the true horses. This is the most specialized of the perissodactyl families; the third toe, which is also the best developed in the rhinoceroses and tapirs, has here become completely dominant. The second and fourth digits are much reduced and only represented by the narrow, elongated splint bones. Horses tread only on the outer edge or tip of the third toe, which is surrounded by a strong hoof. On the inner sides of the legs lie the chestnuts, which are hard, naked, horny thickenings, which probably represent the external openings of vanished skin-glands. The head is large with a small cranium and very powerful jaws and muzzle. The dentition is complete with large incisors, and canines which are well developed in the male but rudimentary in the female. The molar teeth have long crowns and short roots and their biting surfaces are covered with winding enamel ridges,

Quagga

37

which make the teeth well adapted for dealing with the tough grass food. The powerful jaw muscles and the very long gut may be regarded as adaptations for a diet of grass. Hearing and smell are of greater importance for the horse than sight. As a rule the coat is short-haired, the mane is usually erect and the tail has long hairs. Horses live on plains or in other open environments, usually in herds with a leader. When in danger their rapid-moving legs, powerful hearts and spacious lungs allow them to make a quick getaway, but when taken by surprise at close quarters they defend themselves by biting and kicking. The gestation period is about one year and the new-born foal can move about on its own almost immediately. There are differences of opinion about the classification of the horse family; here we are regarding them as all belonging to a single genus, *Equus*, which includes the mountain zebra, Grévy's zebra, the quagga, the different forms of the common or Burchell's zebra, the ass, kiang, kulan, onager and finally the true horses.

The *Zebras* live on the open plains of Africa south of the Sahara, and may be easily recognized by the black or dark-brown stripes on a light skin. The tail ends in a tuft of long hairs, and the ears are intermediate in length between those of the horse and the ass. Chestnuts are found only on the front legs. Zebra herds may contain hundreds of animals, sometimes thousands, which graze freely with antelopes, giraffes and other Ungulates. They often have oxpeckers on their backs, which serve as lookouts and fly off when there is any danger. Nevertheless many zebras are killed by lions, which are certainly their worst enemies. Zebras come down to their drinking-places, usually in the evening, and there the lions lie in wait ready to spring. In captivity they can be crossed with horses and asses, but the resulting *zebroids* are sterile. Zebras are sometimes trained for use in circuses and for driving, but attempts at more extensive training are not successful, and they are inferior in performance to both horses and asses.

Mountain Zebra, *Equus zebra,* ⊥ 49 in. This is the smallest of the zebras and has the most southerly distribution. Once found in enormous herds in Cape Province, but now restricted to a small reserve. The ears are pointed and there is a gridiron pattern of transverse stripes on the hind-quarters.

Grévy's Zebra, *Equus grevyi,* ⊥ 59 in. This is the largest of the zebras. They have numerous narrow black stripes, but no gridiron pattern on the hind quarters. Found in large herds in Abyssinia, Somaliland and northern Kenya. They were unknown to science until 1882, when the Emperor of Abyssinia presented one to the French President, Grévy.

The Quagga, *Equus quagga,* ⊥ 55 in., from South Africa, was exterminated in the 1870's. It had dark stripes only on the head, neck and the front part of the body. The other zebras are best regarded as subspecies or races of the common or Burchell's zebra. These include:

Burchell's Zebra, *Equus burchelli,* from Bechuanaland and Orange Free State, now probably extinct. The back and sides were orange or cream.

Zebras. (top, left) *Mountain Zebra,* (right) *Grévy's Zebra;* (centre, left) *Chapman's Zebra,*
(right) *Burchell's Zebra;* (bottom, right) *Grant's Zebra,* (below) *Boehm's Zebra*

Nubian Wild Ass

and rocky areas of north-east Africa, where they feed on dry leaves and thorny bushes.

The stallions often go about alone except during the breeding season. As a rule asses walk with short steps; sometimes they trot, but seldom gallop.

Nubian Wild Ass, *Equus asinus african-us,* $\perp$ 45 in. The coat is blue-grey or red-grey with a black stripe running along the back, and a transverse stripe over the shoulders. The species is almost extinct in Nubia, Sudan and Eritrea, but is said to occur in certain parts of the Sahara.

Somali Wild Ass, *Equus asinus somaliensis,* $\perp$ 55 in. The coat is red-grey with distinct transverse stripes on the legs. Nowadays they are found only in the barren coastal areas of Somaliland.

Chapman's Zebra, *Equus burchelli antiquorum,* which is found in Bechuanaland, Matabeleland and the Transvaal.

Further to the north there is **Boehm's Zebra,** which is found in Tanganyika and the neighbouring territories. This retains traces of the pale "shadow" stripes between the dark stripes.

Finally there is **Grant's Zebra** in Tanzania and northern Uganda. Chapman's Boehm's and Grant's zebras are commonly seen in zoological gardens.

The *Asses* have long ears, a short erect mane, which starts just behind the ears, a thin tail with a tuft at the end, slender legs with narrow hooves and chestnuts only on the fore-limbs.

Asses are found in the mountainous

Somali Wild Ass

(left, from above) *Giant Ass, Mule, Hinny, Egyptian Donkey;*
(right, from above) *Dwarf Ass, Zebroid*

Kiang

Domestic Ass, *Equus asinus asinus.* A well-known domestic animal, probably derived from Nubian wild ass stock; it occurs in several varieties, varying in size from the **Giant Ass** of Poitou, ⊥ 61 in., to the **Dwarf Asses** of Ceylon and Sardinia, ⊥ 31 in. Domestic asses are mostly found in south-west Asia and in the Mediterranean area. Hybrids be-tween horse and ass have tails like a horse and the ears of an ass. The cross between a male ass and a mare is known as a **Mule,** that between a stallion and a female ass as a **Hinny.** A **Zebroid** is a hybrid produced by an ass and a zebra. All these hybrids are sterile.

We come next to the so-called *Asiatic*

Kulan

Onager

asses, which live in the barren regions of Asia, from Syria to Tibet. They are mostly yellow-brown in colour, with a distinct dark stripe along the back, and they have chestnuts only on the front legs. The ears are shorter than in the true asses but longer than in the true horses.

Kiang, *Equus kiang,* ⊥ 51 in. The upper parts are chestnut-brown, the belly white, and there is a sharp dividing line between the two colours. Kiangs are found in Tibet, where they live chiefly at high altitudes, between 12,000 and 18,000 ft. As a rule they go about in herds of 20-40 animals, but sometimes there may be 300-400, led by an old mare.

Kulan, *Equus hemionus,* ⊥ 47 in. The coat is pale yellow-brown in colour. Kulans are found principally on the steppes of Mongolia, Transbaikal and Transcaspia, in herds which are always led by an old mare. They wander extensively in search of suitable grazing places.

Przewalski's Horse

43

Tarpan

doubt about the purity of the surviving stock. They live in the deserts of Mongolia. They have a short, erect mane without a forelock and a long, hairy tail, which almost reaches the ground in the stallion. They may be fawn, cream or yellowish, but there is evidently considerable variation in coat colour; there is a dark stripe along the back.

Tarpan, *Equus caballus gmelini,* ⊥ 50 in. This horse was first described from the steppes of South Russia about 1770 and had become extinct by 1880. In build they were very similar to Przewalski's horse, but the short, coarse coat was mousegrey in colour. Often several hundred tarpan would stay together, in small herds each with its own stallion leader. They were very watchful and shy, but were not afraid of predatory animals and used to trample on wolves with their front hooves.

Most authorities nowadays consider that the animal described as a tarpan was in fact just a form of the wild Przewalski's horse, dealt with above.

Onager, *Equus onager,* ⊥ 43 in. Onagers are found mainly in Persia, Afghanistan and Pakistan, and go about in large herds led by a stallion. They run so fast that a horseman can overtake them only with difficulty. This is the "wild ass" of the Bible.

The *True Horses.* Our domestic horses originate from tamed wild horses, which are characterized by their large head, short thick neck, heavy body and slender legs; the tail has long hairs, although at its root these only occur on the sides.

Przewalski's Horse, *Equus caballus przewalskii,* ⊥ 53 in. This is said to be the only true horse which still lives in its original state, although there is some

The Domestic Horse, *Equus caballus caballus,* has been living in association with man for some 4,000 years. Without purposeful breeding a number of local varieties have arisen which are not very different from wild horses. The horses of today have been developed partly from these local races and partly by crosses between the local races and the Arab horse.

From the practical point of view horses can be divided into light, "warm-blooded" types and heavier built, "cold-blooded" types; these terms are used to characterize the different temperaments of the horses. Among the more primitive types are the West Norwegian pony, the Icelandic pony and the Shetland pony.

West Norwegian Pony, ⊥ 51-57 in. As a rule these are yellowish in colour, but may be mouse-grey, yellow-grey or yellow-brown. They have a dark stripe along the back, dark legs, a broad neck and an erect mane. They are remarkably well adapted as pack-horses in mountain areas.

The *Icelandic Pony,* ⊥ 47-57 in., is closely related to the Norwegian pony, but is very variable in colour. They are used both for riding and as pack-horses. It has always been the custom to let them fend for themselves both in summer and winter. Since the end of the last century there has been planned breeding of these ponies in Iceland, with a view to producing a stronger animal, retaining the advantages of its race, particularly frugality and endurance, in which it is superior to all the other forms of domestic horse.

Shetland Pony, ⊥ 35-47 in. This is the smallest race of horse in the world. They have a rather broad forehead, a long head and slender legs, and can carry an incredible weight in proportion to their size. Shetland ponies are much used as riding animals for children.

Of the "warm-blooded" races of do-

(back) *Belgian Horse, Jutland Horse;* (front) *Shetland Pony, West Norwegian Pony*

mestic horses we may mention the Arab, the English Fullblood, the Hanoverian, the Lippizaner, the Oldenborger and the Frederiksborger.

The Arab Horse, ⊥ 65-70 in., is considered to be the ideal as regards build, speed and strength. It is characterized by the finely-built head, the slender neck and the way the tail is held high up. Arab horses were probably derived from a tarpan-like stock.

The *English Fullblood Horse,* ⊥ 65-70 in., is descended from our local races crossed with the Arab horse, which was introduced into England in 1602. The title English Fullblood should strictly be used only for horses descended from stock recorded in the General Stud Book, which started in 1791. The term *fullblood* refers to the prominent ap-pearance of the blood vessels under the thin skin, especially after a good gallop.

The *Lippizaner,* ⊥ 63-66 in., is of Andalusian-Arab stock. The Lippizza Stud Farm at Trieste was founded in 1580.

The Oldenborger, ⊥ 69 in., is closely related to the Hanoverian, and has been much used in Denmark during the last fifty years as a driving-horse.

The Frederiksborger, ⊥ 61-67 in., originates from the Frederiksborg Stud Farm in Denmark, which was particularly flourishing during the 18th century. They are nearly all red.

The heavier "cold-blooded" horses are mainly used as carthorses for heavy work.

(back) *Frederiksborger;* (front) *Arab Horse, Oldenborger*

The Belgian Horse, ⊥ 67-70 in. This is one of the world's most typical cart-horses, with a powerful broad back and hairy limbs; it has been developed from the local horse population of Belgium. The coat is usually dapple grey, brown or red.

The Jutland Horse, ⊥ 65-68 in., has been derived from a local population of horses in Denmark. The characters have changed somewhat in the course of time and nowadays it is a medium-sized, well-proportioned, plump carthorse, with shaggy legs; the colour is mainly red.

The *English Carthorses* are similar in many ways to the Belgian, and the finer points of build and performance of these powerful animals belong more appropriately to a book on domestic animals.

Artiodactyls or Even-toed Ungulates

In the artiodactyls, or even-toed ungulates, it is not the number of toes which is characteristic, but the fact that the central axis of the foot runs in between the third and fourth toes, which are more or less mirror images of each other. The second and fifth toes are smaller, do not usually touch the ground when the animal is walking and lie to the side of and somewhat behind the third and fourth toes; the first toe is absent in all living forms. All the artiodactyls, except the camels and llamas, tread on the tips of the toes. Whereas the present-day perissodactyls are only a small remnant of a group which was once much more numerous in species, and seems in fact to be dying out, the

artiodactyls are in the middle of their most flourishing period, and have colonized most parts of the world, with the exception of Australia and New Zealand. The artiodactyls may be divided into two main groups, the non-ruminants and the ruminants. The non-ruminants are those which do not chew the cud, and include the pigs and hippopotamuses. The ruminants, or cud-chewers, include six main groups: the camels and llamas, the mouse-deer or chevrotains, the musk-deer, the true deer, the giraffes and the bovids (cattle, sheep, goats, antelopes etc).

The Non-ruminants

In these the second and fifth toes are usually well-developed, although still much smaller than the third and fourth toes. There are well-developed incisor teeth in the upper jaw and the molars are roughened and adapted for breaking up hard, dry plant food. The stomach is simple in form and not so complicated as in the ruminants.

PIGS

Pigs usually have a compact body, short neck and a short thin tail. The head is wedge-shaped and ends in a snout which carries the nostrils. The upper and lower canine teeth tend to turn outwards and upwards; they grow throughout life and as they lack an enamel layer along one side they become knife-sharp by rubbing against each other, and may thus become dangerous weapons, especially in the boars. The limbs are slender; the central toes are large and powerful and can be spread out, allowing the animal to walk in soft ground. Unlike the other Ungulates the females

47

Wild Boar

nurse a number of young and most have at least eight in each litter.

Wild Boar or **Wild Swine,** *Sus scrofa,* 59+9 ⊥ 35 in. Very variable in size, the large boars may weigh from 330 to 440 lb. The coat consists of long, stiff, dark-brown bristles, with more or less curly, woolly hairs in between. Wild swine go out to feed mainly in the evening and during the night. They live partly on mast and roots, which they dig up with the trunk and front teeth, and partly on worms, larvae and carrion, and may also do damage by eating turnips, corn and potatoes. The breeding season lasts from November to January, and the period of gestation varies from 16-20 weeks. In spring the sow nurses

6-12 pale-yellow striped piglets in a nest of moss and leaves. The young are soon able to move about with the sow; the boar usually goes about alone. Wild boars live in the forests of Central and South Europe and eastwards into Asia. They are extinct in Britain and in northern Europe. During the Stone Age, wild boars were much hunted by man in Europe, and elsewhere.

Apart from this species there are several other kinds of wild pig, particularly in Asia. Amongst these is the *Indian Wild Boar* which has a shorter face but is otherwise closely related to the wild boar; they live in India, Ceylon, Burma and Siam.

Other pigs of this group are the *East Indian Wild Boar,* which has a broad brown band along the middle of the snout, and ranges from Malaya to Java, Sumatra and Borneo, and the *White-whiskered Swine* of China, Japan and Formosa, which has contributed to several of the modern breeds of domestic pigs.

The *Papuan Wild Pig* is found in New Guinea and the neighbouring islands. They are probably descended from semi-domesticated pigs introduced by man. The natives get their tame pigs by catching and rearing wild piglets or by chasing the tame sows out into the forest to pair with wild boars, and then driving them home again.

There is also a tiny pig, known as the *Pigmy Hog,* not much larger than a hare, found in the foothills to the south of the Himalayas.

Domestic Pigs have been derived from wild swine. In Europe during the Middle Ages the tame pigs were not specially cared for; they were allowed to find their food in uncultivated land or in the forests where they searched for mast. Hybridization of these pigs with wild swine from eastern Asia, together with careful selection, has produced the pig breeds which are now found in all advanced agricultural communities; these have a foreshortened head and face, a more or less concave back, a longer body, a curly tail and few hairs. The sow is ready to breed from the age of eight months. She has ten to twelve nipples and usually produces two litters a year, each with about ten piglets. The **Yorkshire Pig** arose from crosses between English pigs and Chinese pigs, and from the 1880's onwards has been much used in hybridization to produce new breeds. At the beginning of the 19th century there were two types of pig in Denmark, the larger Jutland pig and the smaller island pig. From 1896 onwards work began on the improvement of the Danish Landrace pigs and it has been so successful that the **Improved Danish Landrace** is now not only as good as the Yorkshire, but in many respects is perhaps superior.

Red River-Hog, *Potamochoerus porcus,* 51+9 ⊥ 23 in. This handsome pig is found from Liberia to the Belgian Congo. It belongs among the bush-pigs or river-hogs, which live in forest country south of the Sahara or in Madagascar; they have long narrow snouts and pointed ears. They are quite closely related to the wild boar and to our domestic

Pigs. (above) *Yorkshire Pig;* (below) *Danish Landrace Pig*

Red River-hog

pigs, but they have far fewer piglets in a litter. In the red river-hog the bristly hairs of the skin are short and soft, except below the eyes, on the cheeks and on the tips of the ears where they form long tufts. The body is red-brown and the legs and forehead black. The stripe along the back and also those above and below the eye, together with the tufts on the cheeks, are all white. Red river-hogs are lively animals which live in herds in marshy forest country.

The *African Forest-Hog* reaches a length of more than 6 ft. and a weight of over 440 lb. It was first observed in Kenya in 1904 and has since been found to have a wide distribution in the dense forest areas of Central Africa. It has a coat of black bristly hairs and six large warts on the face, of which the two half-moon-shaped outgrowths beneath the eyes are the largest. In the boar the upper canine teeth are the largest and may be more than 11 in. long. Forest-hogs are diurnal animals and feed largely on climbing plants, which they tear

off the trees. They usually live in family groups, well hidden away; the boar defends the sow and piglets so well that they are safe from anything except a lion.

Wart-Hog, *Phacochoerus aethiopicus,* $55+17 \perp 28$ in. Found in savannah country from Senegal to Abyssinia, and further south through East Africa to Natal. The head is very large and there are big warts on the face. In the boar the up-turned canines of the upper jaw may be more than 8-11 in. long. The covering of hairs is sparse, except for a long mane which runs along the back. Wart-hogs usually live in pairs or in small parties, and their food consists of roots, bamboo shoots, fruits, worms and carrion. When rooting in the ground—using the canines more than the snout—they kneel down on the front legs, which have horny thickenings, and push forwards with the back legs.

Babirusa, *Babirussa babyrussa,* $42+8 \perp 31$ in. Found in Celebes and in Boru,

Wart-hog

the westernmost of the Molucca Islands. The canine teeth in the upper jaw grow through the upper lip and then curve upwards and backwards; the canines of the lower jaw grow in the same direction. Babirusas are nocturnal animals, living in marshy, wooded country, where they feed on fruits and larvae. Unlike the other pigs, the sow has only a single pair of nipples and produces only one or two young at a time.

The *Peccaries* are small and grey-black in colour. There are two species, the *Whitelipped Peccary* and the *Collared Peccary;* they are found in South and Central America, and northwards into the southern part of North America.

Babirusa

Common Hippopotamus

HIPPOPOTAMUSES

The hippopotamuses show their relationship with the pigs in their dentition, which has powerful canine teeth; their feet have four toes, all of which touch the ground. Apart from a few hairs along the back and the strong bristles on the lips, they are almost naked. The body is powerfully built, the neck is short and the head large and rectangular.

Common Hippopotamus, *Hippopotamus amphibius,* 160+20 ⊥ 59 in. The males may weigh two or three tons, but females are considerably lighter. The colour grades from copper-red to blue-grey. When a hippo has been out of the water for some time, so that the skin has become dry, it exudes a reddish sweat, which has given rise to the opinion that the animal sweats blood when the skin is dried out; this is not true, as the red colour is not due to a blood pigment. The semicircular canine teeth or tusks on the lower jaw sometimes reach a length of 45 in. and weigh 8 lb. The upper canines grow out downwards and are much shorter; normally they lie within the lips but occasionally one, two or all four canines grow out of the mouth. They are used as ivory, and are finer and harder than elephant tusks. At one time the hippo was common along the whole of the Nile; nowadays it is extinct north of Khartoum, but elsewhere is found in almost all the larger rivers and lakes of tropical Africa, and is most abundant in equatorial Africa from Kenya to Senegal. Hippos are good swimmers and have been seen out at sea off river mouths on several occasions. Inland they are found at heights of 4,000 to 7,000 ft. above sea-level, where it may be so cold in winter that there is ice on the water; in such places the thick skin probably forms a good insulator. By day, hippos usually remain in the rivers and lakes, where they graze on marsh and water plants or lie sunbathing on the mud banks. By night they often go on foraging trips among the plantations, where they eat corn and sugar-cane and destroy the crops by trampling them down. In general they live peacefully in groups of 20-30 animals, but during the breeding season there are terrific fights between the males. The period of gestation is about 8 months, and as soon as the young is born it is led down to the water by the mother, who lies on her side so that it can suckle. The young hippo is somewhat darker than the adult and remains in the care of its mother for over a year; in the water she often carries it on her back. When hippos emerge from the water they shake the water out of their ears, and whilst in the water the males, in particular, shake the tail very fast from side to side, to spread their dung and thus mark out their territory.

Pigmy Hippopotamus, *Choeropsis liberiensis,* 60+6 ⊥ 29 in., weight 550 lb. Compared with the common hippopotamus the skull of the pigmy hippo is relatively larger and the snout smaller, but the upper lips are very large and hang down over the lower ones; the legs are proportionately long. The colouring is a mixture of violet, olive-brown and dark grey. Pigmy hippos are rare, and live, mainly in pairs, in the forest areas of Liberia, Sierra Leone and Guinea. They are not exclusively aquatic, but wander about in the forest at night. The food consist of grass, young shoots and roots. The period of gestation is 7 months; the newly-born young weighs about 13 lb., and remains with the mother for almost 3 years.

Pigmy Hippopotamus

Ruminants

The artiodactyls are divided into two main groups, those which do not ruminate–the pigs and hippos–which we have already described, and the ruminants, which may be further sub-divided into six groups: the camels and llamas, the mouse-deer or chevrotains, the musk-deer, the true deer, the giraffes, and finally the bovids (cattle, sheep, goats, antelopes etc.). In general the ruminants are large animals which are adapted for feeding on plant material. The front (or incisor) teeth in the upper jaw are either lacking or there may be just a single one (on each side) as in the camels; the molar teeth have winding longitudinal ridges of enamel on the biting surface, which help them to crush and grind the food during chewing. The intestine is long and the coecum is large, whilst the stomach is divided into three parts in the camels and into four parts in most of the others. It is characteristic of the ruminants that the food passes through the mouth twice. The stomach is divided into the rumen, reticulum, psalterium and abomasum. When grazing, a ruminant fills the rumen with food. Later on when it starts to chew the cud some of the food in the rumen passes into the next compartment, the reticulum, which contracts and pushes the food up the oesophagus to the mouth. There the food is chewed small and then slips down through the oesophagus, past the reticulum, and through the psalterium to the abomasum where the first part of actual digestion takes place.

Dromedary or Arabian Camel

Bactrian Camel

CAMELS and LLAMAS

The camels and llamas have long curved necks, long cleft upper lips, narrow hind-quarters and feet with only two toes. They stand and walk on two joints of the toes; the outermost joint of each toe has a hoof, which resembles a large flat nail, and behind it is the well-developed pad of the foot. They have an ambling gait, that is, the front and back legs of one side move forward at the same time. This method of progression is more suited for endurance than for speed. Camels and llamas are incredibly easy to satisfy and are content with dry grasses and thorny bushes; they can go without water for days at a time.

Camels

The humps form an important food-reserve, and after long periods of famine they hang loosely to one side or may almost completely disappear. The humps also play an important part in water production, for fat can be burnt completely to water and carbon dioxide, and in this way a hump can produce about its own weight of water, this is surely the main reason why camels can tolerate lack of water for long periods. There are pads of hard skin on the wrist, knee and breast. Camels are peaceful animals and it is only in the breeding season that the peace is disturbed by fights between the males. The

55

gestation period is 11-13 months and the young camel remains with the mother for about a year.

Dromedary or **Arabian Camel,** *Camelus dromedarius,* ⊥ 90 in. There is a single hump, and the colour may vary considerably, but as a rule it is pale yellow-brown. The coat is soft and woolly, and long on the head, shoulders and hump. The male has glands on the back of the head which produce an evil-smelling black secretion during the rutting season. The dromedary is no longer found in the wild state, but as a domesticated animal it is abundant over the whole of North Africa and in a large part of south-west Asia; in recent years it has been introduced to many places in America, Australia and southern Europe.

Bactrian Camel, *Camelus bactrianus,* ⊥ 94 in. This form has two humps and is usually dark brown, with a thick, long-haired coat. A few live wild on the steppes of Central Asia, but most are domesticated and used for riding and as beasts of burden; in addition the wool, hide, flesh and milk are of value. A strong Bactrian camel can cover 18-24 miles a day with a load of 480-580 lb.

Llamas
Llamas have long ears, a short tail and a long woolly coat; they are mountain animals of South America. When excited they can spit, ejecting a mixture of saliva and part of the evil-smelling contents of the stomach.

Guanaco, *Lama glama huanacus,* 80+ 9 ⊥ 45 in. Guanacos are about the same size as a red deer, and weigh about 130-155 lb. The coat is red-brown

Llamas (above); *Guanaco* (below)

56

with white on the belly. They are found in the Andes from Ecuador to Tierra del Fuego and are also common on the plains of Patagonia. They live in herds of females and young animals, led by a male, who keeps watch whilst the herd grazes. Guanacos are hunted for their flesh and hide, and also because sheep-farmers regard them as competitors in areas with little grass.

Llama, *Lama glama glama,* ⊥ 50 in. This is a domesticated form of the guanaco; the coat colour varies considerably and may be black, white, red, orange or mottled. Llamas are kept in Peru and Bolivia, where the males are used as pack animals, the females exclusively for breeding and wool production. They usually cover about 6-12 miles a day when laden; they do not do well in the warmer, low-lying areas near the coast.

Alpaca, *Lama pacos,* ⊥ 43 in. This is another domesticated form of the guanaco; it has a longer and softer coat, which is usually either white or completely black. There are about two million alpacas in the region round Lake Titicaca; they graze in the open throughout the year and are only driven into the towns and villages to be clipped. The wool is made into carpets and cloth in Peru and Bolivia, or is exported to Europe.

Vicugna, *Lama vicugna,* ⊥ 34 in. This is the smallest of the llama group, and it has a reddish-yellow woolly coat. Vicugnas live up in the Andes in Peru, Bolivia and southern Ecuador, and are rare. Carpets and felt hats can be made from their wool, which is also used for knitted goods.

Alpacas (above); *Vicugna* (below)

CHEVROTAINS or MOUSE-DEER

Chevrotains have a ruminant stomach with three compartments like that of the camels, but slender legs similar to those of the true deer. Like the musk-deer they have long canine teeth in the upper jaw, but lack the musk-gland.

The *Water Chevrotain* is dark brown with white markings on the back and white spots and stripes on the neck and sides. They live a sheltered life in the tropical forests of West Africa, from Senegal to the River Congo, feeding by night on leaves and berries; if disturbed they escape into the water and hide.

Malayan Chevrotains. There are several races or subspecies, ranging in size from *Tragulus javanicus*, $27 + 3 \perp 12$ in., to the Kanchil, *Tragulus kanchil*, $18 + 1\frac{1}{2} \perp 8$ in., which is the smallest living Ungulate. The white area on the throat is divided into three bands. They live either alone or in pairs on the outskirts

Musk-Deer

Malayan Chevrotain

of forests and come out at dusk. The period of gestation is 5-6 months and as a rule there are two young. They are found in Malaya, Sumatra and Java.

MUSK-DEER

Musk-deer are sometimes included among the mouse-deer, sometimes among the true deer. They have no antlers, and in the males the upper canine teeth project from the mouth; the males also have a small gland behind the navel, which produces a musky secretion, particularly during the breeding season.

Musk-deer, *Moschus moschiferus,* 38 ⊥ 20 in. In the male the pointed upper canine teeth project about 2 in. from the upper lip and have a sharp back edge; they are used in fights during the breeding season. Musk-deer live in the mountain forests of central and eastern Asia, and northwards to southern Siberia.

DEER
Deer are slender, long-limbed and short-haired ruminants with naked muzzles and pointed hooves. For most of them the characteristic features are the antlers, which are found in all the males, and in reindeer in the females also. Antlers are bony structures carried on the frontal bone of the skull for a part of the year. The lowermost thick part of the shaft is known as the pedicle, which is firmly fixed to and is actually part of the frontal bone. When the deer sheds its antlers there is a bloody wound. The new antlers are at first covered with woolly skin, which is full of blood vessels. Once the antlers are fully formed the blood supply stops, and the woolly skin or "velvet" dries up. The deer then starts to rub the antlers against tree trunks and branches until the "velvet" has been worn off. In the first year of life the antlers are unbranched. After this they acquire more and more branches—or points—as the years go by, but not necessarily one point for each year. The skin of a deer has little or no woolly hair; the summer coat is short and lighter in colour, the winter coat longer, thicker and as a rule darker. Scent glands on the face, just below the eyes, produce a secretion during the breeding season, and there are similar glands between the two hooves of each foot and also on the hind-legs. Deer are shy, nocturnal animals, often living in herds. They usually have one or two young at a time. The deer family is very widespread, but is not found in Australia or in Africa south of the Sahara; their main centre of distribution is in southeast Asia.

Red Deer, *Cervus elaphus,* 68-78+5 ⊥ 47-57 in. The male, or stag, is considerably larger than the female, or hind. The sleek summer coat is red-brown and is replaced in autumn by the longer, thicker and more greyish winter coat; the fur on the hind-quarters "mirror" is always paler in colour. The antlers fall in February-March, and the newly-formed set is fully developed by September. The first antlers have only one point, in the next year there are two points, and in the third year there are usually three points on each antler. The antlers are usually fully developed at an age of 6-7 years. Red deer rub the antlers on trees, like roe-deer, but they do more damage because they strip off more bark. Sometimes they also cause damage to trees during winter by biting with the sharp incisor teeth of the lower jaw. Otherwise they feed on grass, foliage, twigs, buds, heather and mast. Red deer go about in herds and keep mainly to the larger forest stretches with heathland and bog, where they can sun themselves in mud wallows. During autumn the stags fight amongst themselves for the hinds. The victor remains with the hinds, whilst the defeated males move away. In spring the hind bears a white-spotted calf, which acquires the adult colouring in the following autumn. During the Stone Age, red deer were much hunted in Europe. The species is found

Red Deer, stag and hind

spread over almost the whole of Europe and is common in Central Europe.

The *Wapitis* are larger deer, of which there are several races in North America and central and nort-eastern Asia. In North America the wapiti is also known as the elk, a name which is elsewhere used for *Alces alces*, see p. 62.

Axis Deer, *Axis axis,* 55+7 ⊥ 36 in. The body is red-brown with numerous white spots. Axis deer live in India and Ceylon, often in large herds, and they are essentially diurnal animals.

Sika Deer come from Japan and China and have been introduced in cer-

tain parts of Europe. In summer they are chestnut-brown with white spots, in winter they are uniformly dark. The small antlers generally have four points. The period of gestation is about 7½ months.

Père David's Deer is named after a Jesuit priest who first observed this species in the Imperial Park at Peking; it is not known in the wild state. The tail is very long for a deer and the legs rather large and heavy. They were exterminated in China during the Boxer Rebellion in 1900, when the Imperial grounds were raided by the rebels, but have since been reintroduced to the Peking Zoo.

Axis Deer, female, male and female

Fallow Deer, *Dama dama,*
51-59+7 ⊥ 35 in. In summer
the coat is red-brown with
white spots, but individuals are
often found which are either
completely black or completely
white. In its second year the
male, or buck, grows small
antlers with a single point; in
the following year the antlers
are longer and each has 2-3
points, in the fourth year the
uppermost part of each is
spadeshaped, and in the sixth
or seventh year the antlers are
fully developed. The antlers
are shed in spring and the new
ones are fully grown by August-September. Fallow deer

Fallow Deer, female, black buck and white female

feed mainly on grass and are easily satisfied. They live in woodland and herds of them may often be seen in the clearings. The period of gestation is about 8 months, and there is usually only one fawn, which is born in June-July. Fallow deer are native to southern Europe, Asia Minor and northern Palestine, but have been introduced into Britain and many other countries, where they now thrive in a semi-domesticated state.

Reindeer, *Rangifer tarandus,* 65-73+5 ⊥ 43 in. The coat is dark grey-brown in summer, but becomes more whitish-grey in winter, when the hairs may be up to 2½ in. long. This is the only deer in which both sexes have antlers. Reindeer live mainly in treeless mountain regions, feeding in summer on grass and in winter on the lichen known as reindeer moss, which they scrape out of the snow with the broad hooves. They are found in Scandinavia, Finland, Siberia, Spitsbergen, West Greenland and Arctic North America. The Lapps keep them in large herds as domestic animals and use them for transport, milking and meat production. In North America they are known as Caribou.

Elk, *Alces alces,* 110+4 ⊥ 78 in. This is the largest of the world's deer. The coarse coat is dark brown or almost completely black in summer but somewhat lighter in winter. The antlers of the bull elk stand out almost at right angles from the sides of the head and have a broad flat area from which the points spring; they fall in December-April and the new set is complete again in August-September. In the first year each antler has only one point, in the second year two points, and then the number of points increases more or less regularly, until they are fully developed at 10-12 years, when they may weigh 55 lb. Elk live in coniferous or mixed forest, especially in places where there are small lakes and marshes, and feed on leaves and shoots, grass and heather, marsh and water-plants, and may do damage by eating the bark of trees. The cow elk bears a single young the first time, but later usually two; the birth takes place at the beginning of summer after a gestation period of 9 months. They live either alone or in family parties and usually start to look for food immediately after dawn. Elk are found

Reindeer

Elk, cow and bull

in Norway and Sweden, in northern Russia and also in parts of North America, where they are known as Moose. At one time they were more widespread in Europe, for instance they were abundant in the Black Forest in Germany during the time of Julius Caesar.

The *Roe Deer* have small antlers and a very large black snout. They lack the upper canine teeth and the eye glands. Roe deer are found throughout the greater part of Europe and Asia.

European Roe Deer, *Capreolus capreolus,* 39-45 ⊥ 29 in. This is the smallest and commonest deer of Europe, and is found in most wooded areas in Britain, southern Scandinavia and central and southern Europe. In summer the thick coat is short and red-brown, in winter it is longer and grey. The tail is scarcely visible; on the rump there is a large white spot, the mirror, which may serve to guide the kids when they are running after their parents in a dark forest. Smell and hearing are acute, and vision is also well developed. When the male kid is about 6 months old it has very short antler growths, known as buttons, which are shed almost as soon as they have stopped growing. In the next year the young buck grows single-pointed antlers; the maximum number of points in each antler is three. The antlers are shed after the rutting season in Novem-

Chinese Water-deer

Roe Deer. (back) *doe;* (front) *buck*

musk-deer. They live alone or in pairs in the areas of tall grass along the river banks in China and are good swimmers. The female has 4-6 young in each litter, unlike the musk-deer and true deer which have two at the most.

ber-December. At the end of May or beginning of June the doe has usually two kids. Roe deer live in woods and bush country surrounded by fields. They remain hidden during the day, and come out at sunset to search for food in the open country, in summer alone or in families, in winter in small herds. They feed on sprouting grass, winter corn, clover, buds, spruce shoots, heather and moss.

Chinese Water-deer, *Hydropotes inermis,* ⊥ 20 in. These are small deer with no antlers, but with the upper canine teeth protruding below the upper lip, which makes them look rather like

Marsh Deer

Chilean Pudu Deer

den by the hairs on top of the head. They live in dense forest.

GIRAFFES

The members of the giraffe family are restricted to Africa, where they are found only south of the Sahara.

The **Okapi**, *Okapia johnstoni*, ⊥ 62 in., has a short, glossy, dark red-brown coat. The front legs, and the back legs from the root of the tail downwards, are cross-striped with black and white, and there are large white patches on each

Marsh Deer, *Blastocerus dichotomus,* ⊥ 40 in. These are large deer, which live in the marshy forest areas of Brazil, Paraguay, Uruguay and northern Argentina. The antlers are 5-pointed.

Virginian Deer are found throughout the United States of America, east of the Rocky Mountains, and in southern Canada, and there are races of them as far south as Brazil and Peru.

Pudu Deer are America's smallest deer; they have short antlers, each with only one spike. There are two species in the western part of South America.

Chilean Pudu Deer, *Pudu pudu,* ⊥ 13 in. In young males the antlers are hid-

Okapi. (front) *male;* (back) *female*

Reticulated Giraffe

lower leg. The front legs are proportionately shorter than in the giraffe, and not much longer than the back legs. There are two short conical horns on the top of the head in the male only; these are covered with skin, except at the tips which are naked. The okapi is found in the damp parts of the Ituri and Semliki Forests in the Congo and feeds on leaves and shoots. They are shy, nocturnal animals, and are seldom seen. Okapis were not known to science until 1901. At first only a few strips of skin were obtained, which natives used for belts, and as these came from the striped parts on the limbs it was at first thought that the animal was some kind of zebra; the pioneer work in getting hold of specimens of the skin was done by Sir Harry Johnston, after whom the species was eventually named. As soon as complete specimens were examined it became clear that this new mammal was quite a close relative of the giraffe.

The *Giraffes* have been classified in various ways, but here they will be treated as subspecies of a single species, *Giraffa camelopardis,* the tallest and proportionately the shortest of all mammals, which has a shoulder height of 118 in., a total height (to the top of the head) of 195-230 in., a body length of 88 in., and a tail measuring 30 in., excluding the tuft of hairs at the end; on average they weigh 1,100 lb. The front legs are the longest of the limbs, the front part of the body is very high and powerfully built and the neck much elongated. When drinking, a giraffe has to spread its legs out to the side so as to get its muzzle down to the water; exceptionally it may kneel down on the front legs. They live mainly on the leaves of acacia trees, gripping the shoots with the long tongue and then biting them off with the sharp front teeth of the lower jaw. Both sexes have short horns covered with skin; there are always two horns on the top of the head and sometimes a single median one in front of these. Giraffes live on the savannah lands south of the Sahara, from Nigeria to the Orange River, but have been exterminated in the southernmost parts of Africa. They may be divided into a northern and a southern group.

In the *northern giraffes* the coat has brown patches with regular edges on a lighter background, and there is a short bony horn on the front of the head between the eyes. To this group belongs the **Reticulated Giraffe,** *Giraffa camelopardalis reticulata,* from Somaliland and northern Kenya. It has very large dark patches on the skin, with the paler network between them reduced to narrow stripes.

In the *southern giraffes* the brown patches have deeply-indented borders where they join the pale background and the horn on the frontal bone may or may not be developed according to the race. To this group belongs the **Masai Giraffe,** *Giraffa camelopardalis tippelskirchi,* of which both light and dark individuals are found. As a rule giraffes go about in small herds, which consist of an old male, several females and some young males, and they are often seen in company with zebras, antelopes and ostriches. Unlike most other mammals, giraffes have very well-developed sight. Giraffes are often said to be dumb, but in fact they do occasionally make rather weak bleating noises. The gestation period is 14-14½ months.

HOLLOW-HORNED RUMINANTS

These form the last and by far the largest of the six families of ruminants. The horns are unbranched and consist of a horny outer sheath, which corresponds with the horny layer of the skin, and which surrounds a bony core firmly fixed to the frontal bone of the skull. The horns, which as a rule are carried by both sexes, keep growing and are not shed, except in cattle, where the calf's first small horns are lost and replaced by the permanent ones. This common character is not, however, shared by the prongbucks. The hollow-horned ruminants have no incisor or canine teeth in the upper jaw, whilst in the lower jaw the canines have moved up close to the incisors and have the same chisel-like shape. The six molar teeth in each half of the jaw have very high crowns. Usually they have only one young every year. Most of the species live in Africa, and there are none at all in Australia or South America. They may be divided into the pronghorns, antelopes, sheep, goats, musk-ox and cattle.

Pronghorn

PRONGHORNS to a certain extent form an intermediate group between the deer and the hollow-horned ruminants. The forked horns occur only in the male; the horny sheath is shed every year, leaving the unbranched core naked. Hairy skin then grows over the core and forms a new thin horny sheath.

Pronghorn, *Antilocapra americana,* 51 +7 ⊥ 35 in. The coat is smooth and yellow-red in colour, with large areas of white, and there is a rather short mane along the back of the head and on the upper part of the neck. The fully developed horns measure 9-11 in. and become forked when the male is 3 years old. They live on the prairies and in the neighbouring forest areas of North America. Reckless hunting has decimated the former large herds, but the species is now strictly protected. Mating occurs in September-October and the two young are born in May.

ANTELOPES

The *Wood Antelopes* differ from the other antelopes in that they live in forest areas. They occur in Africa south of the Sahara, with a single species in India.

Bushbuck or **Harnessed Antelopes,** *Tragelaphus scriptus,* ⊥ 30 in., have a mane along the whole of the back; horns occur in the male only and may be 7-11 in. long. The coat colour varies considerably according to the race. They are common in forest areas from Senegal to Angola.

Nyala, *Tragelaphus angasi*, ⊥ 42 in. The males have long horns, up to 27 in., and long hair on the neck and chest. They live in south-east Africa from Malawi and Rhodesia to Natal.

Greater Kudu, *Strepsiceros strepsiceros*, 94+23 ⊥ 52 in. The horns of the male are twisted and very long, and may measure 40 in. in a straight line. They are widely distributed in Africa, from Abyssinia to the Cape.

The *Lesser Kudu*, ⊥ 40 in., has no mane on the breast. It is found from Abyssinia southwards to the region of Tanzania.

Bushbuck or Harnessed Antelope.
(front) *male;* (back) *female*

(left) *Nyala, male and female;* (right) *Greater Kudu, male and* (front) *female*

The *Elands* have a large head, a much-folded dewlap and a tufted tail. The horns are long and stout in the male, but even longer and more slender in the female. Elands have long hairs on the forehead and a short neck mane.

Eland, *Taurotragus oryx,* 110+24 ⊥ 40 in. Elands live more in open country than do the other antelopes in this group. They are found in Kenya, Tanganyika and the Kalahari Desert. On the slopes of Kilimanjaro they may go up to an altitude of 12,000 ft. The herds are usually led by an old male. In spite of weighing some 2,000 lb., the Eland is a good jumper. The period of gestation is about 9 months.

The Giant Elands from Senegambia and the Sudan are the largest of all the

Nilgai, male and female

Eland, male and female

antelopes. They have broader ears than the common eland and a black neck; they live more in forest country.

Nilgai, *Boselaphus tragocamelus,* 75+ 19 ⊥ 54 in. The males have long (7-11 in.) straight horns and a dark steel-grey coat with a short mane on the neck and a tuft on the throat. The females are hornless and have a grey-brown coat. Nilgai live in small herds in India from the Himalayas to Mysore, but they are rare in the northern part of their range. The female usually has two calves at a time, after a gestation period of 8 months.

The *Reedbucks* are about the size of a roe-deer, have a short woolly tail and a naked gland-spot below the long ears. Only the males have horns.

Reedbuck, male and female

brown coat, and is white round the eyes, snout and ears; there is a distinctive white elliptical band round the tail and rump. Common waterbuck live in the forests of East Africa and are excellent swimmers. During the mating season the male gathers a whole harem of females around him, and will often climb up a termite colony in order to keep a good eye on them.

Lechwe Waterbuck, *Kobus leche,* $\perp$ 40 in. A smaller animal than the common waterbuck, and found in the marshy areas of Nyasaland, Rhodesia and Angola.

The *Horse-like Antelopes* are large, powerful animals with long, ringed horns in both sexes. The head has a dark band on the forehead and snout,

Reedbuck, *Redunca arundinum,* 58+ 11 $\perp$ 36 in. The horns of the male measure 15-16 in. Reedbuck live on grassy plains of Africa south of the Sahara, and they are never far from water. They are fast-running, agile animals, most commonly found in places with tall grasses.

The *Grey Rhebok* is found in South Africa south of the Zambesi. It lives in open hilly regions, coming down into the valleys at night for water.

Waterbuck are found everywhere in Africa south of the Sahara. Only the males have horns.

The *Common Waterbuck* is about the size of a red deer and has long horns and a long tail; the coat has coarse greasy hairs; the long tail has a tuft of hairs at the end. This species has a grey-

Lechwe Waterbuck.
(back) *male,* (front) *female*

71

Sable Antelope (left), *and Roan Antelope*

and the long tail ends in a tuft. The sub-family may be divided into three groups: the roan and sable antelopes, the oryx and the addax.

Roan Antelope, *Hippotragus equinus,* 88+28 ⊥ 6o in. This is a handsome animal with an erect mane and very large ears with tufts at the ends. As a rule the horns are about 30 in. long. Roan antelopes live in bush country from Senegal to Abyssinia, and southwards to the Orange River.

Sable Antelope, *Hippotragus niger,* 88 +28 ⊥ 54 in. The long horns are bent backwards and outwards, and are usually 40 in. long, but may exceptionally be over 6o in. long. The coat of the male is shining black or dark brown, that of the females more brownish. The ears are about 9 inches long and have no tuft at the end. Sable antelopes are distributed from Angola to Tanzania and southwards to the Orange River, usually in open hill country.

The Blue Buck was a related antelope found in Cape Province; the last specimen was shot about 18oo.

Oryx have very long, straight or slightly curved horns, very short ears and no mane. They are found in Syria, in the deserts of Arabia and on the plains of Africa south of the Sahara. They use the long, pointed horns as weapons of defence and are not so shy as other antelopes. They often go about in herds together with the sharp-sighted springbucks, and seem to be able to manage for long periods without water.

Beisa Oryx, *Oryx beisa,* 80 + 15 ⊥ 47 in., are found in open country in Somaliland, Kenya and Tanganyika. Their numbers have been much reduced by reckless hunting.

The *South African Oryx,* the *Gemsbok* of the Boers, is closely related to the Beisa oryx, from which it may be distinguished by the large black patch on its thighs. Gemsbok once lived in South Africa from southern Angola to the Cape, but have now been exterminated in the areas south of the Kalahari Desert.

White or **Scimitar Oryx,** *Oryx algazel,* 82 + 15 ⊥ 40 in. Unlike the other oryx the horns are curved backwards. The coat is whitish-yellow with rust-red markings on the face, neck, shoulders and hind-quarters. They live on the deserts and plains of Africa from the Sudan to Senegal. They appear to wander extensively, either in pairs or in small herds. The period of gestation is 8-8½ months.

Addax Antelope, *Addax nasomaculatus,* 70 + 11 ⊥ 40 in. In this species the horns are twisted, and in front of them there is a tuft of long black hairs, which hangs down over the forehead. The broad hooves help to prevent the animal from sinking into the sand. Addax live in the desert regions between the Nile and Lake Chad, usually in small herds; nowadays they are much persecuted by gun-carrying Bedouin, and have already been exterminated in some parts of their original range.

The *Hartebeests* and *Gnus* are large animals with broad snouts and very long skulls; both sexes have horns. They are found in large herds on the African

(from above) *White or Scimitar Oryx, Beisa Oryx, Addax Antelope*

73

(l-r) *Korrigum, Konzoni or Lichtenstein's Hartebeest, Blesbok*

plains, often in company with zebras, gazelles and ostriches. The sub-family has three genera, containing the so-called bastard hartebeests, the true hartebeests and the gnus.

The *Bastard Hartebeests* are those which differ least from the other antelopes. They were once found in enormous masses, mostly on the plains of South Africa, but they now survive only in smaller herds.

Korrigum, *Damaliscus korrigum,* $\perp$ 49 in. Found in grassy country from Senegambia eastwards to Uganda and the borders of Abyssinia.

Blesbok, *Damaliscus albifrons,* $\perp$ 35 in. The white face-blaze is divided by a brown bar running between the eyes. Blesbok used to be extremely abundant but are now restricted to a few protected areas.

The *True Hartebeests* are large, rather clumsy animals with long heads and muzzles; both sexes carry horns. There are several different species spread over Africa.

The *Kongoni* has a sandy-yellow coat, and is one of the commonest antelopes In East Africa, ranging from Tanganyika to the Tana River on the Equator.

Konzoni or **Lichtenstein's Hartebeest,** *Bubalis lichtensteini,* $\perp$ 48 in. The remarkable shape of the horns is responsible for the fact that fighting males often get so locked together that it is almost impossible for them to separate.

74

Konzonis are found in East Africa from Tanzania southwards to Malawi and Rhodesia.

The *Gnus* differ considerably from all the other antelopes; they rather resemble a horse with the legs of an antelope and the head of an ox. The very large head has long bristly hairs, a broad muzzle with large nostrils, small eyes and powerful horns (in both sexes); there is a prominent mane on the back of the head and neck. They move with an ambling gait. The peculiar voice sounds rather like a series of hiccups. As a rule gnus live in herds of cows, calves and young animals, led by an old bull.

White-bearded Gnu, *Connochaetes taurinus albojubatus,* 97+25 ⊥ 52 in.

Found in Kenya and Tanganyika, and usually regarded as a race of the brindled gnu.

Brindled Gnu, *Connochaetes taurinus,* 97+25 ⊥ 52 in. The coat is grey with dark cross-stripes; the beard on the chin and neck is black, and so are the mane and tail. Found to the south of the white-bearded gnu and down to the Orange River.

White-tailed Gnu, *Connochaetes gnu,* 92+23 ⊥ 44 in. At one time white-tailed gnus were found in enormous herds in South Africa, mostly around the Vaal and Orange Rivers, but like so many other hunted animals they have now been exterminated in the wild and only a few survive in a couple of large reserves in the Orange Free State.

(l-r) *White-bearded Gnu, Brindled Gnu, White-tailed Gnu*

Klipspringers. This group has only one species with a number of races.

Klipspringer Antelope, *Oreotragus oreotragus,* 39 ⊥ 21 in. Unlike other antelopes, the klipspringer is adapted for life in rocky mountainous areas. There are short straight horns in the males only. Klipspringers are found in rocky regions of Africa from Eritrea and Abyssinia to the Cape, often at altitudes up to 7,000 ft.

The *Duikers.* These are very small antelopes with a large tuft of hairs on the top of the head which almost hides the small horns. They are found in almost the whole of Africa south of the Sahara, mostly in dense forest areas.

Yellow-backed Duiker, *Cephalophus sylvicultrix,* ⊥ 32 in. This is about the size of a roe-deer and has a broad straw-yellow band on the back. It occurs mainly in south-west Africa.

Red-flanked Duiker, *Cephalophus rufilatus,* ⊥ 10 in. The body is not much larger than that of a hare and the back is greyish. They live in the tropical forests of West Africa.

Dik-diks are small antelopes with an elongated almost snout-like muzzle. The females are larger than the males. They occur chiefly on the plains of East Africa.

Dwarf Antelopes. These are small antelopes from equatorial Africa.

Royal Antelope, *Neotragus pygmaeus,* ⊥ 10 in. This is the world's smallest antelope; it has very small horns. Found in western Africa.

(from above) *Klipspringer Antelope, Yellow-backed Duiker, Royal Antelope, Red-flanked Duiker*

Grant's Gazelle and Dama Gazelle

The *Four-horned Antelope* of India is a small animal in which the males have two pairs of horns; the females have none.

The *Gazelle-like Antelopes*. A large group of long-legged animals with ringed horns and short tails, found in Africa and parts of Asia.

The *True Gazelles* are elegant antelopes with lyre-shaped horns. They are found in north and east Africa and in west and central Asia.

Grant's Gazelle, *Gazella granti*, ⊥ 34 in., is found in Kenya and northern Tanganyika. The horns are longer than in any other gazelle.
Thomson's Gazelle, also from Kenya and Tanganyika, has a prominent dark band along the side of the body.

Dorcas Gazelle has a pale, red-brown coat with a band on the side which is not så dark as in Thomson's gazelle. They are found in north Africa from Morocco and Algeria to the southern boundary of the Sahara.

Dama Gazelle, *Gazella dama*, ⊥ 36 in. The largest of the gazelles; it lives in the area between the Sudan and Senegal.

Blackbuck, *Antilope cervicapra*, 50+5 ⊥ 31 in. Blackbuck live on the open plains of India, sometimes in very large herds. They are incredibly fast and can only be hunted with cheetahs. The old males are dark grey-brown or almost black, with white on the snout, around the eyes, on the belly and on the inside of the legs; the females and the young

Blackbuck, female and male

Gerenuk, male and female (left); Springbok (right)

animals are much paler than the male. Only the males have horns.

Gerenuk, *Lithocranius walleri,* 50+11 ⊥ 39 in. An elegant animal with a very long neck and slender legs. The lyre-shaped horns are found only in the males. Gerenuks live in the driest areas of southern Abyssinia, Somaliland and Kenya. They feed on the leaves and shoots of acacia trees; to reach these the animals sometimes stand on the hind legs with the front limbs resting on the branches.

The *Springboks* have a fold of skin along the rearmost part of the back which they can spread out so that its white colour covers the back; scent glands open out at the bottom of this fold. Springboks often jump up six feet into the air in order to get a view while they are running. Both sexes have horns.

Springbok, *Antidorcas marsupialis,* 52 +7 ⊥ 32 in., were formerly found in enormous herds in South Africa south of the Zambesi, but nowadays they live only in small numbers in and around the Kalahari Desert. When their pastures fail they undertake long migrations.

The *Impala* has S-shaped horns which may be up to 32 in. long. It is distributed over eastern and southern Africa.

The *Saiga Antelopes* are somewhat plump animals with thick swollen snouts.

Saiga Antelope, *Saiga tatarica,* 46+4 ⊥ 30 in. The coat is thick and woolly, which makes the hornless females look rather like sheep. The horns of the males are almost lyre-shaped and have

Saiga Antelope, female and male

pale translucent sheaths. Saigas are shy and nervous animals which go about in large herds. In May the female has one or two young, which remain with her until the next mating season. Saigas are found in Asia on the steppes of western Siberia, Turkestan, Dzungaria and Mongolia, and in Europe on the steppes between the Don and the Volga. During and after the Ice Age, when most of Europe had a steppe climate, saigas lived as far west as England and France.

Chiru or **Orongo,** *Pantholops hodgsoni,* ⊥ 31 in. The snout is not so swollen as in the saiga, but the horns of the male are considerably longer. The chiru has a thick woolly pelt and lives mainly in Tibet and eastern Turkestan at altitudes of 12,000–18,000 ft. During the mating season the males fight vigorously for the females, but for the remainder of the year these shy animals live together peacefully in small herds.

Chiru or Orongo, male and female

The *Goat-Antelopes* are hollow-horned ruminants which are intermediate in many respects between the true antelopes on the one hand and the sheep and goats on the other. Like the sheep and goats they are essentially mountain animals and are found principally at high altitudes in south-east Asia; in addition there is one species in Central and South Europe and one in North America.

Grey Himalayan Goral, *Nemorhaedus goral,* ⊥ 27 in. The bushy tail is relatively long and the horns are about 7-8 in. in length. The coat is long with rather erect hairs. They are found in Kashmir and the western Himalaya, often at altitudes of 3,000–9,000 ft.

Grey Himalayan Goral

Serow

Chamois

Serow, *Capricornis sumatriensis,* ⊥ 38 in. The body is reddish-brown or almost black, with a long greyish or reddish-grey mane, long ears and 9-in. long horns. They are found from the Himalayas in Nepal, where they live at heights of 6,000–9,000 ft., southwards to Malaya and Sumatra. In the Himalayas the kids are born in May–June, but in Burma not before the end of September. Closely related forms are found in Tibet and China.

Chamois, *Rupicapra rupicapra,* 42 + 1½ ⊥ 29 in. In the chamois the horns spring straight out of the top of the head and bend back sharply towards the tips; in the male the horns may be 9½ in. long, but they are shorter and thinner in the female. Behind the horns there is a groove with a pair of rutting glands, which become much swollen in the male during the mating season, when they give off a rank smell. In summer the coat is short, in winter long and thick, with the hairs ending in black tips, and almost long enough to form a mane on the front part of the back. It is this tuft of hairs which the hunters wear in their hats. The chamois is found in most of the mountain chains of central and southern Europe, particularly in the mountains of Cantabria, the Pyrenees, the Appenines, the Alps, the Carpathians, in part of the Balkans including Mount Olympus, and in the Caucasus. In Switzerland the chamois had almost died out at the end of the last century, but extensive protection has allowed the population to recover. The chamois is a woodland antelope frequenting the high forest belt and in summer extending up to altitudes of 14,000 ft. When a herd is grazing one or more animals are always placed as lookouts; the warning signal is a piercing note followed by stamping with one of the front feet. Chamois are wonderful climbers and can jump 20 ft. They live on grass, foliage and fresh shoots of trees and bushes. The rutting season is in November–December, and after a gestation period of 6 months the female bears one or two kids, occasionally three. The stomach sometimes contains hard masses, which are composed of resin, hairs and splinters of wood all matted together.

Takin, *Budorcas taxicolor,* ⊥ 39 in. This is a powerfull-built animal with horns which resemble those of a gnu. Takins live in the eastern part of the Himalayas, mostly at altitudes of 6,000–9,000 ft., and are essentially forest animals, frequenting rhododendron scrub and bamboo thickets. In summer they gather into large herds, which break up again at the beginning of spring.

Rocky Mountain Goat, *Oreamnos americanus,* 51 ⊥ 39 in. The pelt is completely white, and consists of long fine woolly hairs and stiff guard hairs. Behind the short, slightly curved horns there are rutting glands corresponding to those in the chamois. Rocky Mountain goats are certainly the most goat-like of the antelopes. They are found in the mountains of north-western North America from Alaska to Montana, living either in pairs or in small herds; in summer they may go up as high as 12,000 ft.

SHEEP and GOATS. This sub-family consists of medium-sized animals with hairy muzzles, cleft upper lips and horizontal pupils in the eyes; the horns are triangular or pear-shaped in cross-section. All the wild species are mountain animals.

Takin

Rocky Mountain Goat

Mouflon, male and female

Argali, male and female

The *Sheep* have spirally twisted horns with transverse rings; the legs are proportionately long. They live wild in the northern hemisphere and may extend above the tree-line, sometimes up to 18,000 ft., where the only other mammals that can thrive are certain goats and the yak. In summer they feed on grass and other plants and in winter they make do with twigs, bark and dead grass. After a gestation period of 5–6 months the female gives birth to one or two lambs, exceptionally to three or four. The lambs grow fast and are able to breed when only one year old.

Mouflon, *Ovis mouflon,* $43+4 \perp 26$ in. The dark horns may measure up to 32 in. in the males; those of the female are only about $2–2\frac{1}{2}$ in. long. Mouflon live in forests along the tree-line, where they remain hidden by day, only coming out to feed at dusk. In autumn the males fight for the females. The flocks are led by an old ewe, but when they are resting an old ram keeps watch. Mouflon are found wild in Corsica and Sardinia, but their numbers are much reduced. There are, however, introduced stocks in several places in central Europe.

The Armenian Wild Sheep is found in Asia Minor, Armenia and Persia, and a closely-related form, the *Cyprian Wild Sheep,* in Cyprus.

Argali, *Ovis ammon,* $78+4 \perp 47$ in. The largest of the wild sheep, with horns which may reach a length of 74 in. Found in central Asia.

The Bighorn Sheep are all but extinct, but are still found in the national parks of North America.

Sheep. (top) *East Friesian Milk Sheep;* (centre, left) *Merino Sheep,* (right) *German Heath Sheep;* (bottom left) *Black-headed Somali Sheep,* (right) *Caracul or Persian Sheep*

83

Domestic Sheep, *Ovis aries,* are found in many different races and, due to interference by man, have become so different from the original wild stock that they are no longer able to survive on their own. Mating may take place throughout the summer, but most farmers arrange for it to take place as late as possible, so that the lambs are born in the spring. Usually there is only one lamb, but twins are not rare. The ewe lambs are sexually mature at 12 months, the males at 18 months. Domestic sheep now have a very wide distribution, being most abundant in Australia, New Zealand, Argentina and South Africa. They reached America at the time of Columbus, but were not brought to Australia until 1788. The thick woolly coat is a character which has been mainly developed in the course of domestication; other selected characteristics are the smallish horns, which may be absent altogether, the floppy ears and the great variation in colour. Domestic sheep have probably been derived from various wild species. They are among the most ancient of our domesticated animals. Among the different kinds of domestic sheep the following may be specially mentioned:

Merino Sheep, which probably originated in Asia Minor, whence they came to Greece and later to Spain. In the centuries up to the 1700's they were bred and reared only in Spain. They are renowned for their fine, curly, thick wool.

The *Marsh Sheep* are hornless and live in the coastal regions of Europe from Schleswig-Holstein to Western France. One of the races is the **East Friesian Milk Sheep,** ⊥ 31-35 in., which weighs 130-220 lb.; it has a large udder and gives a milk very rich in fat.

The **German Heath Sheep,** ⊥ 21-23 in., weighs only 22-33 lb. and is found principally on Lüneburg Heath in Germany. It is usually dark with a black head.

The **Black-headed Somali Sheep** belongs to the *Fat-tailed Sheep,* in which the tail has large deposits of fat and may weigh 20 lb. Fat-tailed sheep are found in Africa, western Asia and southern Europe.

Barbary Wild Sheep. (back) *female;* (front) *male*

Caracul or **Persian Sheep** are small dark-grey forms with large drooping ears and mostly with horns. The coat is long and coarse. The fur known as Persian lamb is obtained from lambs which are killed at an age of 3-8 days. "Broadtail lamb" is taken from unborn lambs.

Barbary Wild Sheep or **Arui**, *Ammotragus lervia*, 62+9 ⊥ 39 in., are found in the mountains of North Africa. In the male the horns are up to about 31 in. long, in the female only 15 in. The mane is particularly long in the male; it grows on the underside of the neck and right down to the front legs.

The Bharal or *Blue Wild Sheep* is a separate species found in Tibet. The coat is blue-grey in colour, and the ram has long powerful horns. It forms a transition between the sheep and the goats.

The *Goats* have a powerful body, strong legs and a short neck; both sexes usually have horns. The coat consists of coarse guard hairs and fine wool hairs, and as a rule the females have a large beard. The males have a characteristic rank smell, particularly during the mating season. Goats are mountain animals and excellent climbers. Mating takes place late in the autumn and after a gestation period of 5 months the female gives birth to one or two kids.

East Caucasian Tur, *Capra caucasica*, 58+5 ⊥ 35 in., are found in herds in the Caucasus. The beard is short and the horns almost completely smooth, with the tips turning inwards.

The *Ibexes* are wild goats which live

East Caucasian Tur, male and female

Alpine Ibex, male and female

85

Wild Goat. (back) *female;* (front) *male*

high up in the mountains of Europe, Asia and North Africa.

Alpine Ibex, *Capra ibex,* 55+5 ⊥ 35 in., were once common in the Alps, but have been much hunted. Nowadays they are protected and are still found in a few places in the Alps, from the tree line up to the glacier zone. During the mating season, from December to January, the males fight strenuously, using their horns, which are up to 44 in. long, and have strongly marked transverse ridges.

Wild Goat, *Capra aegagrus,* 46+7 ⊥ 37 in. The horns are very long and compressed, with sharp front and back

edges. Wild goats are variable in colour, and are found in a number of races in western Asia from the Caucasus to Baluchistan, and also in Crete, from altitudes of 4,500 ft. upwards. In winter they live in large herds, which may sometimes contain 40-50 animals. In April-May the females give birth to 1-2 or sometimes 3 kids. Like the chamois their stomachs often contain smooth, shiny concretions; these were at one time thought to be valuable as a drug, and were used amongst other things as an antidote to poisoning. These "stones" may vary in size from a pea to an apple, and are coloured green, brown, grey or black; they consist of several layers of resin and hair, usually laid down round a pebble or other foreign body.

Tahr, *Hemitragus jemalhicus,* 70+3½ ⊥ 39 in. The curved horns have a

Tahr. (back) *female;* (front) *male*

86

sharp front edge and may be 11-15 in. long in the male, but are much shorter in the female. The coat is thick with long guard hairs and in the male there is a long mane covering the front part of the body. The fur is usually dark brown, but may sometimes be pale slate-grey with a rust-red sheen on the flanks. They are found on the south side of the central Himalayas. They live in herds.

The *Markhors* are large goats found in Afghanistan, Kashmir and Baluchistan. The spirally twisted horns of the male may be 39-58 in. long. The fur is longer than in other wild goats.

Domestic Goats, *Capra hircus*, occur in several different races and are among the most ancient of domesticated animals. They are capable of breeding at an age of 6 months and the females usually give birth to 1-2 kids at a time. The domestic goat goes feral very easily, and this has happened amongst other places, on some of the isles of Greece and on St. Helena, Mauritius and Juan Fernandez. They may be divided into milk-goats and fur-goats, or into goats with drooping ears and erect ears, or into those with and those without horns. Among them are:

The **Saanen Goat**, ⊥ 31-35 in., found in the Saanen Valley in Switzerland. It has a white coat, and a large prominent beard, but lacks horns. The udder is large and the milk rich in fat.

Domestic Goats.
(from above) *Saanen Goat, Roe-deer-coloured Goat, Walliser Goat, Angora Goat*

87

The **Roe-deer-coloured Goat,** which is similar to the Saanen goat but is pale brown and more slender.

The **Walliser Goat,** which is black and white; the males having long, spirally twisted horns.

The **Angora Goat,** which is distributed throughout the whole of western and central Asia. The coat consists of long, soft, silky wool hairs, and is usually white, although it may sometimes be yellowish-grey or black. The wool is known in the trade as mohair.

MUSK-OX. Musk-ox are ruminants which have characters in common with sheep, cattle, gnus and takin, but perhaps most resemble big long-haired rams. There is only one species.

Musk-ox, *Ovibos moschatus,* 97 ⊥ 50 in. The coat is heavier and warmer than that of any other animal; it consists of very thick woolly hairs mixed with long curly guard hairs, which may be up to 23-27 in. long and reach right down to the hooves; only the mane on the throat has straight hairs. The horns particularly in the male, are very broad at the base, where they meet over the forehead; they curve downwards along the side of the head, then outwards, and finally forwards and upwards. As a rule musk-ox go around in small herds and feed mainly on grass, moss and arctic willow; in summer they graze night and day, only stopping to chew the cud. During the mating season in July-September the bulls have a strong musky smell and they have terrific fights with each other over the females; in these fights they rush at each other and strike

Musk-ox

with their foreheads, but do not actually use the tips of the horns. The gestation period is 9 months and the cow usually bears a single calf every second year. When attacked by Arctic wolves, sledgedogs or man, musk-ox form into a circle, facing outwards, with the calves in the middle. Musk-ox are only found in the most northerly parts of Canada and in north-east Greenland, where they are now protected. They were introduced into Norway in 1924 and 1932, and were probably exterminated there by the Germans during the Second World War, but have since been reintroduced.

CATTLE. The last sub-family of the large family of hollow-horned ruminants –the cattle–consists of large heavily built animals. Apart from a few forms, which have no horns, both sexes carry smooth, round, outward-facing horns. The snout is broad with a large naked muzzle, with nostrils opening out to the sides and a divided upper lip. They have a long tufted tail and usually a dewlap on the throat. The female, or cow, has udders with four nipples. The coat is usually short and smooth, and there is sometimes a mane. Cattle occur wild in Europe, Africa, Central and South Asia and North America. As a rule they keep to a walk, but most are capable of a good trot, and when necessary they can break into a gallop. When threatened or attacked they rush forward with lowered head, toss the attacker up in the air on their horns before trampling him underfoot; this may happen to Carnivores or to man. Of the senses, smell is the best developed, then hearing, whilst their sight is relatively poor. A herd of cattle is usually quiet and peaceful except during the breeding season when the bulls fight amongst each other. The period of gestation is 9-12 months and the cow normally has only one calf at a time. Cattle may be divided into three groups–the anoas, the buffalos and the true cattle.

Anoa, *Anoa depressicornis,* 66+11 ⊥ 39 in. Found only in the mountain regions of Celebes. They are the dwarfs among the cattle and in size, coat marking and horns they rather resemble antelopes. Anoas live in the forests, take to the water willingly and have a musky smell.

Anoa

(left, back) *Domestic Buffalo,* (front) *Kerabau;* (right) *Indian Buffalo*

Indian Buffalo, or **Arna,** *Bubalus bubalis,* ⊥ 72 in. Distributed throughout India, Ceylon, south-east Asia, Java, Sumatra and Borneo. Like other buffalo, they like to spend a lot of time in the water or lying in the sun. They feed mainly on water-plants.

Domestic Buffalos, ⊥ 52-66 in., have been derived from the Indian Buffalo. They are found throughout southern Asia, in Africa only in Egypt, and in Europe in southern Italy, Hungary, the Balkans and South Russia. In Southeast Asia and the Philippines there is a race known as the **Kerabau.**

The Cape Buffalo from South Africa is a quite separate species with horns which are very thick at the base. Several races are found over the rest of Africa south of the Sahara.

Gaur, *Bos frontalis gaurus,* 117 + 30 ⊥ 70 in. The bull has a long low hump on the front part of the back. Gaur are found in India, Burma and Malaya, and feed on grass and bamboo shoots.

Gayal, *Bos frontalis frontalis,* ⊥ 66 in. The broad forehead carries thick conical horns and there is a hump on the back similar to that of the gaur. Gayals are found in Burma east of the River Brahmaputra and northwards up to the Himalayas. They appear to be a domesticated form of the gaur.

Bantin, *Bos banteng,* 82 + 35 ⊥ 60 in.

(from above) *Gaur, Bantin, Gayal, Bali Bantin*

The lips, the lower part of the legs and the stern are usually white. Bantin are found in India and the East Indies, and occur both wild and domesticated. Domestication was originally undertaken on Bali, whence **Bali Cow** meat is exported.

The *Bisons* have short, broad, three-cornered heads and short round horns; the front part of the body has a big hump with a mane.

European Bison or Wisent

European Bison, or **Wisent,** *Bos bonasus,* $113+23 \mp 68$ in. In spring the thick winter coat moults and comes off in large tufts. The cow is considerably smaller than the bull. Wisent are forest animals and feed on grass and branches. At one time they were found all over Central Europe, and the Polish princes and Russian tsars kept enormous areas of forest where they hunted them. A hundred years ago there were still some 2,000 head of wisent in Central Europe. The population then decreased and during the First World War there was no chance to protect them. After the war hunger among the human population led to their being further decimated and the remainder were destroyed by poachers. In 1918 there were 200 wisent left, three years later only three, and on 9 February 1921 a former forest ranger shot the last one on Bialowicz Heath in Poland. In the virgin forests of Kuban in the Caucasus there were about 600 head before the war, but here also the post-war period was decisive. The tanners offered high prices for the skins and this sealed the animals' fate. Destruction was carried out with machine-guns and by 1925 the population was reduced to 25. Some years later Europe's largest land mammal had been exterminated in the wild. There were, however, some 28 wisent left in zoos and the stock was gradually increased to about double this number. During the Second World War the wisent population was again in danger, but they survived and there are now over 200 wisent, most of them in Bialowicz and in the Caucasus.

American Bison, *Bos bison,* $117+25 \perp 68$ in. In North America this species is known as the buffalo. They have larger and broader heads than the wisent,

American Bison

Yak, *Bos grunniens,* 125 + 29 ⊥ 64 in. The horns are round, and in the male they may reach a length of 31-35 in. The black or dark-brown coat is very long, so long on the belly that it may touch the ground. Yak are found in the high parts of Tibet at altitudes of 12,000 to 20,000 feet. The bulls usually live alone, but the cows, juveniles and calves gather in herds which may often number some hundreds. Yak are good clim-

Yak

thicker horns and a higher and more developed mane on the back. Some hundreds of years ago millions of them grazed on the prairie lands of North America. They formed the most important quarry of the Indian hunters, but by the turn of the last century they had been almost exterminated by the white settlers. With protection the population has recovered and there are now several thousands in the National Parks of North America.

bers and are the only cattle which live in mountainous regions. During the mating season in September-October the bulls are active throughout the twenty-four hours and run round grunting after the cows. As a rule the cows produce a single calf every second year. The Tibetans use domesticated yaks for riding, as beasts of burden and for the production of meat and milk. These domesticated yaks are smaller and very variable in colour, and many of them are hornless; they are found far beyond the borders of Tibet.

Auroch, *Bos primigenius,* ⊥ 74 in. This fine species of ox is now extinct, but from early descriptions and the study of its skeletal remains we know quite a lot about its appearance and habits. The back was straight, the head rather long and the large horns were directed forwards and had black tips. In the old bulls the smooth, short-haired coat was black with a white stripe along the back and a white muzzle, whilst the cows and the young animals were brown and the calves red-brown. Aurochs lived in small herds in the forests and used to feed in the evening and at night. They were widely distributed in Europe, Asia and North Africa. Their remains have been found particularly in southern Sweden, Denmark and northern Germany, but they lived throughout more or less the whole of Europe, except in the northernmost regions. In Asia they lived in Turkey, Iraq, Siberia and China, and in North Africa in Algeria and Egypt. Aurochs were much hunted and the blame for their

disappearance must fall almost entirely on man; the last one died in 1627 in Jaktorowska Forest near Warsaw.

Domestic Cattle, *Bos taurus,* consist of several races or varieties, which may all be regarded as originating from the auroch, and which now form man's most important domestic animal in most parts of the world. They are used both for milk and meat production, and to a certain extent as transport and draught animals; the bulls can be used as stud animals when they are 1-1½ years old, but it is not until they are 3 years old and fully grown that they can mate with up to 3 cows daily, with a maximum of 150 cows in a year. The period of

Auroch

(from above)
Jutland Bull, Red Danish Milking Cow, Ankole Cattle, Gujarot Zebu, Nellore Zebu

gestation lasts 285 days. The new-born calf is taken from the cow immediately after birth. Normally a cow produces milk only whilst the calf is small, but she will go on producing milk for a considerable time as long as she is milked. The calves are fed with full milk during their first month and then weaned on to skimmed milk. Most of the bull calves are slaughtered, but some of them are castrated and used as draught animals. Before the birth of her first calf the female is known as a heifer. The age at which a heifer first calves varies from two to over three years, depending on the race, and on the amount of feeding. Domestic cows come on heat at intervals of about three weeks and remain thus for 1-2 days. When a cow has calved, she comes on heat again after 4-6 weeks.

Locally produced races of cattle have been established in a variety of different natural conditions, and they may be black, grey, brown, yellow, white or mottled, with considerable variations in size. Selection within a race and crossing with other races has produced the more highly valued varieties. Some of these have been developed for meat production, others for milk, but most of them are dual-purpose cattle. In Britain there are several meat races, such as Shorthorns and Herefords. To the milk-producing races, which are the most numerous, belong the Friesians, **Red Danish Milking Cow,** the **Jutland Milker,** and the Jerseys and Guernseys. Jerseys are a carefully developed race of milk cow, which produce top quality milk with a high fat content.

Ankole Cattle are found on the Masai Plains and along Lake Victoria and Lake Tanganyika. They have larger horns than any other cattle—over 50 in. in length and 19 in. in circumference at the base.

Zebu have elongated heads, large dewlaps and a large hump on the shoulders. They are found from central India eastwards into China, westwards to Egypt and the Sudan, and southwards as far as South Africa. The Hindus regard the zebu as a sacred animal, which is often allowed to wander free in their villages. The largest race is the **Gujarot Zebu** from Pakistan, which has long horns. In the **Nellore Zebu,** on the other hand, the body and the horns are somewhat smaller.

Elephants

THE Elephants are the world's largest living land mammals. They are usually classified near to the Hyraxes and to the Sirenians (manatee and dugong) and not far from the Ungulates. At first sight these four mammal groups do not appear to have much in common except that they are all vegetarian, but a study of extinct, fossil forms has shown that they all have common ancestors.

The most characteristic features of the elephants are their limbs, trunk and teeth. The legs are long and each one has five toes which have large elastic pads. There are 3-5 hooves on each foot. The elephants are amblers and seldom break into a trot or a gallop; when walking ordinarily they do about 2-4 miles per hour, but when frightened they can move off at the rate of 35-40 miles per hour.

The upper lip and snout are drawn out into a long trunk with the nostrils at the end. The trunk functions as a kind of universal tool. Thus it is a well-developed organ of smell, which is often held up in the air like a periscope, it is also a tactile organ of great sensitivity, particularly at the tip; as a gripping organ it can fetch down branches and twigs or pull up whole trees by the root; it can be filled with water, which is then squirted into the mouth, or with sand, which is shot over the back to remove insects; and finally it can function as a weapon in the form of an outsize rubber truncheon.

Elephants have no canine teeth, but they do have a single pair of incisor (or front) teeth in the upper jaw—the tusks—which project from the mouth and continue to grow throughout life. At first there is a little enamel at the tip of each tusk, but this quickly wears off and for the remainder of their lives the tusks lack any trace of enamel. As a rule, elephants only use one tusk at a time as a working tooth; when this one becomes sufficiently worn, the other acts as a substitute, whilst the first grows out to the same length again. These large and elegant tusks are the source of ivory, and many thousands of elephants have been killed by man in his search for this valuable animal product. The molar teeth have broad, flat upper surfaces with enamel-capped cross-ridges. Usually only one molar tooth in each half of each jaw is in use at a time; as it becomes worn a new tooth emerges behind and eventually takes its place. The front end of the new tooth comes into use whilst its hind end is still hidden in the jaw.

The colossal head contains a mass of air-filled cavities in the space between the brain and the outer surface. The brain itself weighs about 11 lb., that is, about four

Indian Elephant. (left) *female;* (right) *male;* (front) *young*

times as much as that of a man; it is, however, a relatively small brain, for it is only about $1/1000$ the weight of the body, whereas that of man is $1/45$. The question of intelligence in elephants has often been argued, but there is no doubt that they have a good memory.

The eyes are small and vision is poor, but hearing and smell are both acute. The ears are very large, in the African elephant so large that they seem to cover almost a sixth of the whole body surface. They function not only in hearing, but also in temperature regulation, for the skin behind the ears and on their back surfaces is very thin; when elephants stand and flap their ears they are helping to keep their temperature down. When the ears are held outwards they help to give the impression that the animal is even larger than it really is.

Elephants live exclusively on plant food and get through an enormous quantity of leaves, bark, branches, fruits, grass and roots–at least 100-200 lb. per day, and they may drink 20-40 gallons of water in the same period. They spend almost sixteen hours a day in searching for food and eating it, and up to two hours in polishing the tusks with dry leaves; on the other hand they only sleep for two to four hours in the day.

Elephants go around in herds searching for food and water, with an old cow as

African Elephant, female (left), *and male*

leader and with the males bringing up the rear; but when there is any question of danger they change round so that the bulls are in front.

The period of gestation is 22 months, which is longer than in any other mammal, and the cow elephant suckles the young one for almost two years; her two nipples are placed forwards near the front legs. The longevity of elephants is often exaggerated; they seldom live more than 60 years, and the trained working elephants are usually worn out at an age of 50 years.

As regards training it is peculiar that the elephant is perhaps the only animal which can be trained equally well whether it be captured as an adult or taught from birth. We are inclined to think that only Indian elephants can be trained, but this is not so, even though nowadays very few African elephants are trained. In Ancient Egypt elephants captured in Abyssinia were used by the Ptolemies for military purposes, and in the First Punic War the Romans overran a Carthaginian camp and captured 120 elephants. Hannibal started with 50 elephants when he invaded Italy, but lost most of them during his crossing of the Alps. Before this time Alexander the Great brought back many Indian elephants from his campaign in India. But this was nothing compared with the numbers kept by the Indian and Persian rulers, for the King of Persia is said at one time to have had an army of 9,000 elephants.

Indian Elephant, *Elephas maximus,* 10½ ft. tall in the male, 8 ft. in the female; weight 4-5 tons. Found in the tropical forests of India, Ceylon, Sumatra and Borneo. The forehead is steep and somewhat concave, with protruding frontal knobs, and the trunk ends in a single finger-like process. The tusks of the bulls may be 8 ft. long (record 8¾ ft.) and weigh 44 lb.; the cows either lack tusks or have very small ones. The front legs have five hooves and the back legs have four.

African Elephant, *Loxodonta africana,* 11½ ft. tall, weight 5-6 tons. Found in the African savannah country south of the Sahara. The head has a flat sloping and convex profile and the trunk ends in a pair of finger-like processes. In the bull the tusks may measure more than 10 ft. (record 11 ft. 5½ in.) and weigh more than 110 lb.; the tusks of the cows are considerably smaller. The ears are enormous. The front legs have four hooves, and the back legs three.

In addition to the ordinary form of African elephant there is the smaller *Forest Elephant* which is found in the forests of West Africa; this should perhaps be regarded as a subspecies. Other races have been distinguished in the size and shape of the ears, but these are not very reliable characters. African elephants are tamed and used as working animals in the Congo.

Sirenians

THE Sirenians or Sea Cows are at first sight rather similar in appearance to the whales; the body is streamlined and almost hairless, the fore-limbs are paddle-shaped and there are no external hind-limbs. But these are only characters which might be expected in mammals living in the water. In actual fact it is more correct to regard the sea cows as elephants which have taken to the water, and the two groups have much in common; for example they have the same kind of tooth succession for the molar teeth, and in both groups the nipples are situated between the fore limbs. Sirenians feed on seaweeds and other aquatic plants which they find along the coasts of tropical seas and in river mouths. At high tide they may swim up into lakes and marshy areas; when the tide goes out they are often stranded there and killed by the natives who eat their flesh.

The eyes are small and there are no external ears, but the muzzle is large with crescent-shaped nostrils which are open only when the animal is breathing. The thick lips have long sensory bristles. Sirenians graze at night in shallow water where they only have to lift up the head in order to breathe. By day they rest and sleep with the nostrils just above the surface; they usually breathe a couple of times in succession and can then remain quite long under water, probably over 16 minutes. When suckling the young the femals holds it between her front limbs. These rather peculiar animals are said to have given rise to stories of mermaids, although some authorities

Manatee

Australian Dugong

consider that the latter are more likely to have been seals. The Sirenians are divided into three families: the manatees, the dugongs and Steller's sea cow

Manatees

The manatees have a rounded tail fluke, very sparse hair and thick bristles round the snout. There are four species, one from tropical West Africa and Lake Chad, one from the Amazon, one from Florida and one from the West Indies.

West Indian Manatee, *Trichechus manatus,* about 8-9 ft. long. Found in the Bay of Mexico and northern South America; they are very common in the Orinoco.

Dugongs

The dugongs have a slightly forked tail fluke and even less hair than the manatees. There are two incisor teeth in the upper jaw which grow into small tusks in the male. There are three species (or perhaps only races) of dugong, one in the Red Sea, one along the coasts of the Indian Ocean and a third around Australia and New Guinea.

Australian Dugong, *Halicore australe,* 160 in. These are found only on the coast and do not go up into the rivers. They live in small herds.

Sea Cows

Steller's Sea Cow, 20-25 ft. long, was first observed in the Bering Sea in 1741, but in the course of about thirty years was almost completely exterminated by whalers and sealers. They are believed to have become finally extinct in the middle of the 19th century.

Hyraxes

THE Hyraxes form a small independent order of mammals, containing about ten rabbit-sized species. In Leviticus XI, where they are called conies, we read of the cud-chewing cloven-footed animals which the Israelites may or may not eat. "And the coney, because he cheweth the cud, but divideth not the hoof; he is unclean unto you". The Hebrew text uses the word "saphan", which means a coney or hyrax. The passage is not, however, correct from the zoological point of view, for in fact the hyraxes do not chew the cud. It is possible that the characteristic twitching movement of the hyraxes was interpreted as a form of cud-chewing. Nevertheless the israelites were probably wise to forbid the eating of hyrax flesh, for the alimentary canal usually contains a heavy load of tapeworms and roundworms.

At one time the Hyraxes were included with hares and rabbits among the Rodents, partly because they are about the same size and partly because their front teeth are somewhat similar to those of the Rodents. There was also a time when they were thought to be related to the Insectivores and Ungulates but they are now classified in a separate order close to the Elephants.

Hyraxes live in colonies in most parts of Africa south of the Sahara, and in the near East. They are not often seen, but they leave evidence of their presence in the form of very considerable piles of excrement which are deposited near their holes. At one time this dung was much sought after as a cure–for internal use–for hysterics and anaemia.

Hyraxes have four toes on the front limbs and three on the back limbs; the soles of the feet are covered with firm pads which are used in obtaining a good grip on trees and rocks. The toes all have nails, except the second toe on each hind foot which has a curved claw. Hyraxes climb with great agility up the perpendicular sides of rocks, and they are also remarkable jumpers, thinking nothing of jumping down from a height of 9-15 ft. From perpendicular or even overhanging cliff faces 24-30 ft. high they come down like a cat, sliding or running three-quarters of the way before taking off with all four legs and landing elegantly on the ground.

The fur of a hyrax is very soft—even though it consists only of guard hairs. In recent years it has appeared on the market as a solid and cheap fur, useful amongst other things for fur linings. There is a narrow pale patch of hairs on the middle of the back; these hairs can be erected to form a circle round an area of naked skin, which has several scent glands. These glands produce an aromatic secretion which probably plays a part in bringing the two sexes together during the mating season.

Another peculiar feature of the fur is the presence of long protruding bristly hairs; these occur not only round the snout and the eyes, but also in rows down the back and along the limbs. It is probable that these hairs have a sensory function. The fur in general is usually infested with fleas and lice.

Hyraxes have two or three well-developed young in a litter.

Abyssinian Hyrax

Tree Hyrax, *Dendrohyrax dorsalis,* 17½ in. Found in the tropical forests of Africa. Tree hyraxes sleep by day in hollow trees and come out at night in search of fruits, leaves and buds.

Abyssinian Hyrax, *Procavia habessinica,* 17½ in. Found living gregariously in Abyssinia at heights up to 9,000 ft. As soon as they have more or less denuded an area of food, flocks of hyraxes will move on to new grazing-grounds.

Tree Hyrax

Whales

THE Whales are the only mammals (apart from the Sirenians) which are adapted to live exclusively in the water. The body is tapered at both ends, and the head, trunk and tail merge gradually into each other. The powerful muscular tail ends in a broad *horizontal* fluke which is moved up and down very fast when the animal is swimming; in fishes the body ends in a *vertical* fin. The skin is smooth and almost devoid of hairs, and it covers a thick layer of fat, known as the blubber, which may be up to 15 in. thick.

The fore-limbs are in the form of flippers, which are rigid along their whole length, except at the shoulder joint; they are used for steering and balancing. The hind-limbs are lacking externally and are present in some whales only in the form of bony vestiges in the muscles under the skin. The head is very large and often makes up a third of the whole animal. The eyes are very small and are placed near the corners of the mouth; the ear openings are almost invisible and are closed with ear-wax, which acts as a protection against increases in pressure when the animal dives. The nostril or nostrils are on the top of the head, usually well behind the snout. The diaphragm runs obliquely from the belly side up towards the back, thus leaving plenty of space for the lungs on the dorsal (or back) side of the body. The windpipe runs direct from the lungs to the nostrils. The frequency of breathing is very variable; sometimes a whale breathes in 50-60 times at short intervals and then exhales, before diving again, but often it may breathe only 4-5 times. The spout of water, often many feet high, which whales shoot up into the air is derived mainly from the layer of water which was above the head when the whale surfaced to breathe; if the air temperature is low, the spout may also contain damp breath, which has condensed to water vapour.

Although they have no vocal chords whales can produce sounds, and they can also send out ultrasonic waves and thus get to know whether there is any echo-producing prey or enemy in the vicinity. As a rule whales only have one young at a time, which attaches itself to one of the two nipples, which are situated in a pit on each side of the anus. The young whale cannot suckle in the same way as the young of land mammals, because under the water it cannot produce the vacuum which is necessary for the milk to be sucked in. The milk produced by the mother collects in large reservoirs and the contraction of certain of the body muscles causes this fatty milk to be ejected forcibly into the mouth of the youngster.

Susu or Gangetic Dolphin

Whales can sometimes remain underwater for more than an hour, but if caught fast in fishing-nets, and therefore unable to reach the surface to breathe, they will naturally drown. When stranded, the chest region is crushed by their own weight and they quickly suffocate.

Whales are found in all the seas of the world and there are a few in rivers. Usually they do not go out far from the coast, for it is there that they most easily find their food in the form of fish, squid or plankton. Most whales make long migrations every year. The period of gestation is usually 10-12 months, but may perhaps exceptionally be 16 months. They can reach an age of over 50 years.

Whale-hunting has gone on for over a thousand years, and in certain areas there has been over-fishing which accounts for the fact that many of the large species are now very rare. The main aim of whale-fisheries is, of course, the production of whale-oil, which is used in the manufacture of margarine and other products.

The various species of whales are classified into two quite distinct sub-orders, the *toothed whales,* which may be large or small, and the *whalebone whales,* which are nearly all giants. The porpoise with a length of 4¾ ft. is the smallest toothed whale, and the blue whale, 100 ft. long, is the largest whalebone whale.

Toothed Whales

The toothed whales usually have many teeth–sometimes over 200–which are almost all similar in shape and size and which cannot be divided into incisors, canines and molars, as in other mammals. Toothed whales feed mainly on fish and squid. There is only a single nostril or blowhole, which opens to the outside quite far back on the head.

There are five families of toothed whales: the freshwater dolphins, the dolphins, the white whale and narwhal, the sperm whales and the bottle-nosed whales.

Freshwater Dolphins

The freshwater dolphins are regarded as primitive forms; it is in fact thought that whales are descended from land mammals which went to live first in fresh waters and later moved out into the sea.

Susu or **Gangetic Dolphin,** *Platanista gangetica,* 8 ft. Found in the rivers Indus and Ganges. When undisturbed they surface at 30-40-second intervals to breathe. They feed mainly on fishes and freshwater shrimps and are much hunted by the Indians.

In addition to the Gangetic dolphin there are two freshwater dolphins in South America (the *Amazonian Dolphin* and the *La Plata Dolphin*) and one in China (the *Chinese River Dolphin*).

Dolphins

Common Porpoise, *Phocaena phocaena,* 4-6 ft. Found in the coastal regions of the North Atlantic and neighbouring seas, and on the American side of the Pacific Ocean; it is common in British waters. The flesh was once appreciated in Britain and the oil was used for lamps. Porpoises go about in pairs or in small schools and usually surface to breathe 2 or 3 times in succession, and then remain submerged for 3 or 4 minutes. They have 23-27 small teeth in each half jaw and feed mainly on her-ring, small cod and salmon, although they also take crustaceans and squid. The young is born about June and is almost half the length of the mother at birth; it grows very fast, which is not surprising for the milk contains 40-50% fat.

Common Dolphin, *Delphinus delphis,* 7-8 ft. The back is dark brown, the belly white and the sides yellowish-grey; the snout is long and there are 40-50 teeth in each half of the jaw. Common dolphins usually go about in schools, hunting for pilchards and other small fishes; they often follow ships and can be seen jumping out of the water. They are widespread in all the warmer seas and are common in the warmer parts of the Atlantic Ocean, Mediterranean Sea and Black Sea, and are frequently seen along the south coast of Britain.

The Killer Whale, from Arctic and temperate seas, is the sea's fastest swimmer and most feared predator. It can kill porpoises and seals, and small schools of killer whales can overcome whale-bone whales which they attack and bite in the corners of the mouth or on the tongue.

Common Dolphin and Common Porpoise

(above) *Sperm Whale or Cachalot;* (below, left) *Bottle-nosed Whale,* (right) *Pilot Whale*

Pilot Whale, *Clobicephala melaena,* 28 ft. Recognizable by the blunt forehead and the long narrow flippers. Pilot whales live almost exclusively on squid and usually wander about in schools of up to two thousand animals in the cold and temperate seas of the northern and southern hemispheres. In the Faeroe Islands one or more schools are driven ashore every year, either into a bay or into shallow water where the animals are killed with long knives.

White Whale and Narwhal

The *White Whale,* 12-15 ft., is completely white and has no back fin. It is an Arctic whale, and feeds mostly on fish.

The *Narwhal,* 12-16 ft. (excluding the tusk), is greyish in colour. The male has a spirally twisted tusk, about 8 ft. long, growing out from the upper jaw. Narwhals are found in Arctic seas, often in large schools. They feed mainly on squid, but also take fish and crustaceans.

Sperm Whales

Sperm Whale or **Cachalot,** *Physeter catodon,* 40-60 ft. This is the largest of the toothed whales. The colossal head contains a mass of tissue filled with spermaceti oil, which solidifies to a soft white mass when the animals dies. Spermaceti is used in the manufacture of candles and cosmetics. Big lumps of ambergris are sometimes found in the gut; this is a solid, greyish substance, which fetches high prices and is still used in the preparation of the more expensive scents. The upper jaw usually lacks teeth, but there are 20-27 large cone-shaped teeth in the lower jaw. The food consists almost entirely of large squids. Sperm whales can dive to a depth of over 3,000 ft. and remain submerged for more than an hour; when they return to the surface they will often breathe more than 50 times before diving again. They are found in small schools in all seas, especially in the warmer parts.

Bottle-nosed Whale, *Hyperoodon ampulatus,* 25-32 ft. These have a short snout in front of the fatty pad on the head. There are one or two teeth in the front part of each half of the lower jaw; they feed mainly on squid. Bottle-nosed whales are found in small schools in the Atlantic and they move southwards in winter.

Baleen Whales

The baleen whales are toothless, but they have triangular horny plates of baleen which are fixed crosswise on the roof of the mouth. These plates are closely packed one behind the other, and their inner edges are frayed into a series of fringes which form a kind of sieve. Baleen plates may be regarded as enormously developed folds of the palate.

Baleen whales live principally on small crustaceans and other animals in the plankton, which they separate from the water by means of the baleen sieve.

Blue Whale

There are two families: the fin whales which have a dorsal fin and the right whales in which this fin is lacking.

Fin Whales

Blue Whale, *Balaenoptera musculus,* 80-100 ft. This is the largest of all whales and also the largest animal that has ever existed: a blue whale may weigh over 150 tons. They live almost exclusively on krill–shrimp-like crustaceans only 1-2 in. in length. Blue whales spend the summer in temperate, Arctic and Antarctic seas and in winter move into the sub-tropics. They can swim at a speed of 18-24 miles per hour. Every second year the females give birth to a single young whale which is 20-24 ft. long at birth; it is suckled for the first 6 months. During this period is grows at the rate of 1½-2 in. in length and 220 lb. in weight every day; the milk contains 35% of fat. When the young whale is one year old it measures about 50 ft., and it is fully grown at two years.

The *Piked Whale* or *Lesser Rorqual,* 23-33 ft., feeds on small shoaling fish. It is found in the Atlantic and often comes quite close to the coast.

The *Humpback Whale,* 32-58 ft., has characteristic knobs on the upper side of the head and along the edge of the very long flippers.

Right Whales

North Atlantic Right Whale, *Balaena glacialis,* 45-59 ft. The mouth has about 250 pairs of baleen plates, each one about 12 ft. long.

The *Greenland Right Whale,* 50-60 ft. has a gigantic head which makes up nearly a third of the total length. The mouth contains over 300 pairs of baleen plates, which are even longer than those of the North Atlantic right whale. The Greenland right whale has the most northerly distribution of all the baleen whales; it occurs along the edges of the Arctic icefields and is thought to be almost extinct.

North Atlantic Right Whale

Seals

THE Seals are really a group of Carnivores specialised for life in the sea. Their limbs are short and directed backwards and all four feet are webbed. On the fore-limbs the digits decrease in length as one passes from the first to the fifth, but on the back limbs the two outer digits are longer and stronger than the three middle toes. The nostrils and ear-holes are closed when the animal is diving. Like all air-breathing aquatic mammals, the seals have to come up to the surface to breathe at frequent intervals, although exceptionally they can remain submerged for several minutes. When asleep in the water they rise to breathe at regular intervals and then sink down again. There is a thick layer of fat or blubber beneath the skin.

Seals feed exclusively on other animals, and mainly on fish. Apart from man they have no enemies except the polar bear and the killer whale. The group may be divided into the true seals, the eared seals and the walruses.

True Seals

The true seals have a short neck and no outer ears to protect the ear passage, but the ear-holes close when they are submerged. The greater part of each limb is hidden within the body; the feet stick out, are hairy on both sides and have a web between the toes which extends a little beyond the longest toes. When swimming, seals hold the hind-feet vertical with the soles close to each other, so that together they somewhat resemble the tail fin of a fish. They swim by moving the back part of the body and the hind-limbs from side to side and only now and again use the weak front limbs for propulsion. When hunting fish, the head is withdrawn towards the shoulders and moved forwards again only to seize the prey. On land the limbs are unable to support the body or to move it forwards, but seals manage to move about clumsily by hunching up the back and then throwing the body forwards with the front limbs held close in to the sides.

Common Seal, *Phoca vitulina*, 59-72 in. Very common on both sides of the North Atlantic and the commonest seal in British waters, where it breeds on small islands and rocky reefs; there is a closely related form in the North Pacific. The pups are born on land at the end of June, usually on sandbanks or rocky beaches; they moult their woolly baby-coat in the first few days; this allows them to go into the water almost immediately. The adult seals weigh 150-220 lb., the new-born cub 25 lb.; during its first ten days, when it is being fed only by the mother, it puts on several pounds weight per day, for seal milk contains nearly 50% fat (ordinary cow's milk has 3.4%). Later on the young feed on prawns, but the adults live almost exclusively on fish, such as herring, cod, whiting, flounders, and salmon; they also eat whelks. On an average they eat 11 lb. of fish per day and this, together with the fact that they tear the fishermen's nets, makes them unpopular among fishing communities; in some countries rewards have been paid for dead seals in an effort to reduce their numbers.

(from above) *Ringed Seal, Grey Seal, Common Seal*

Ringed Seal, *Phoca hispida,* 39-63 in. This is the smallest of all the seals. The adults have a pale ring round the dark spots on the skin. Ringed seals are found in the North Polar seas and are common off Greenland, but they also occur as a relict from the Ice Age in the Baltic Sea and the Gulf of Bothnia, whence they occasionally wander into Danish waters. Closely related seals live in inland waters, such as Lake Ladoga and the Caspian Sea. The pups are born out on the ice during the spring and they suckle from the mother for 3-4 weeks. The adults feed mainly on fish and crustaceans. Only occasional specimens have been seen in British waters.

Harp or **Greenland Seal,** *Phoca groenlandica,* 71-78 in. These are animals of the high Arctic, found far out from the coast on the drift-ice in the Arctic Ocean, often in enormous flocks. The young are born on the ice. Enormous numbers of both young and adults are killed annually for their skins; the blubber is also of some value.

Grey Seal, *Halichoerus grypus,* 78-98 in. Found on both sides of the Atlantic Ocean, and in the North and Baltic Seas. There are now estimated to be about 50,000 grey seals in these areas; 5,000 off the coasts of Canada, 3,000 round Iceland and the Faeroes, 20,000 round the British Isles, 400 off Norway and over 5,000 in the Baltic. On the coasts of Britain the young are born in the autumn and during their first weeks they have a long, soft silky coat and do not enter the water. The food consists of fish and crabs.

Leopard Seal, *Hydrurga leptonyx,* 120-140 in. Found mostly on the drift-ice in

Crab-eating Seal (back); *Leopard Seal* (front)

the Antarctic, but also extending north-wards to New Zealand and Australia. Leopard seals (or sea-leopards) are fast and capable swimmers, which are able to jump high up out of the water on to ice-floes. They feed principally on cuttle-fish, but also catch penguins and other sea birds.

Crab-eating Seals, *Lobodon carcino-phaga,* are the palest in colour of the seals. They are mainly Antarctic in dis-tribution, keeping almost exclusively out among the drift-ice, where they feed on crustaceans.

Bladder-nosed Seal, *Cystophora cristata,* 78-98 in. Bladder-nosed or hooded seals live in the Arctic Ocean, mainly out on the drift-ice, and seldom come in to the coasts. The adult male can inflate a sac on the top of the head, which is in com-munication with the nostrils, to form the hood or bladder. They feed on fish and squid.

(left, front) *Bladder-nosed Seal, male,* (back) *female;* (right, back) *Greenland Seal, female*
(centre) *male,* (front) *pup*

Northern Elephant Seal

Patagonian or Giant Sea Lion

The *Elephant Seals* are the giants among the seals; they get their name from the trunk-like snout which is developed in old males, and which may reach a length of 30 in. The bulls are almost twice as long as the females, and weigh over 3 tons. The thick layer of blubber may make up about a third of the total weight.

Northern Elephant Seal, *Macrorhinus angustirostris,* 97-200 in. Now found only near Guadaloupe Island, off Lower California, where there is a colony of about 1,000 animals. During the breeding season they spend 3 months ashore, but the rest of the time they undertake extensive migrations. They feed on squid and fish.

The Southern Elephant Seal lives, amongst other places, in South Georgia and on Kerguelen Island, and is much commoner than the northern elephant seal. Otherwise the two species are similar in appearance and habits.

Eared Seals

These can be distinguished from the other seals by the small external ears. In many respects they have not diverged so far from land mammals as the other seals. They have a longer neck, the feet are naked on the underside, and can be turned inwards under the body, so that they can move about quite easily on land and even climb up rocky slopes. The webbed skin of the back feet extends beyond the tips of the toes and is strengthened by special cartilages. The front limbs are also used on land, but are particularly used in slow swimming; when swimming fast the hind-limbs are also brought into play. The claws are only fully developed on the three middle toes of the hind-limbs.

The eared seals are usually divided into two groups: the sea lions and the fur seals. In the sea lions there is no undercoat of woolly hairs, such as is usually found in seal pups, whereas in fur seals there is a thick under-coat of fine woolly hairs throughout life. The males are always much larger than the females.

The *Northern* or *Steller's Sea Lion,* 117-136 in., lives in the northern part of the Pacific, and as far south as southern California.

The Californian Sea Lion, 78-87 in., is the best-known of the sea lions, being the one usually seen in zoos and circuses, where it is much admired for its outstanding sense of balance. They live along the coast of California southwards to Mexico and are sometimes found in enormous herds.

They are excellent swimmers and can leap up 3 ft. out of the water. In captivity they are usually fed on herrings, but in nature, fish forms only a part of their diet, which consists mainly of squid. The pups are usually born in June and live ashore for the first 6-8 weeks, feeding entirely on the mother's milk.

Patagonian or **Giant Sea Lion,** *Otaria bryonia,* 97-117 in. The old bulls have a thick mane round their neck and often roar like a lion. They live along the coasts of Patagonia, Chile and the Falkland Islands and are very similar in habits to the other sea lions.

Northern Fur Seal, *Callorhinus alascanus,* 50-72 in. Found in enormous numbers on the Pribilov Islands. The bulls have a short mane. Fur seals come ashore in spring to fight strenuously for the cows, and old bulls may collect a harem of 40-50 cows, which they guard assiduously. The pups are born in the middle of summer, and a couple of days after the birth the adults mate again, either on land or in the water. During their first four weeks the pups are scarcely able to swim, and it is not really until the autumn that they enter the water freely. Fur seals feed on herring, sar-

Northern Fur Seal. (left) *female;* (right) *male;* (front) *pup*

Walrus

dines, salmon and other fish, as well as on squid, The fur is known in the trade as "Alaska Seal"; to prepare it for market the stiff guard hairs are removed by loosening the hair roots from the underside, leaving only the soft, reddish woolly hairs; fashion dictates that these woolly hairs should be coloured chestnut-brown or coal-black. Some hundreds of years ago there were millions of fur seals, but wasteful slaughter of hundreds of thousands every year brought the stock almost to the point of extermination around 1910. Partial protection has, however, saved them, and it is now estimated that there are about 1½ million Northern Fur Seals on the Pribilov Islands, and from this stock 50,000 skins are harvested every year by the United States Government.

Walruses

Atlantic Walrus, *Odobenus rosmarus,* 117-175 in. Like the eared seals, walruses can stand up on their limbs, which are naked on the underside, but they lack the external ears. The tusks are formed by the canine teeth of the upper jaw, and they are used as weapons and also probably for digging molluscs up from the sea-bottom; the other teeth are small and flat and well adapted for crushing the shells. They also eat young seals. Walruses have a thick layer of blubber and a tough naked hide. They live gregariously in the Arctic Ocean, particularly on the drift-ice off the coasts of Greenland, Spitsbergen and Labrador. They are now rather rare, and the population of West Greenland has been estimated at scarcely more than 8,000 animals. Walruses are occasionally seen off northern Norway, the Orkney and Shetland Islands and North Scotland.

The closely related Pacific Walrus, *Odobenus divergens,* is found in the Bering Sea. Its tusks are somewhat longer, up to 30 in., but otherwise it is very similar in habits and appearance to the Atlantic Walrus.

Carnivores

ALL mammals which feed on other live animals are in effect carnivorous, for example bats and insectivores, but in the systematic sense the term Carnivores is used for those mammals which are included in the cat, hyaena, civet, weasel, dog, raccoon and bear families, and also for the seals and sea lions which have been describel in the previous chapter.

The Carnivores are mostly animals of medium size, which feed principally on other mammals, and on birds, fish, insects and other animals, but some of them also eat berries, juicy roots and other vegetables material. They have their main area of distribution in the tropics. They walk either on the whole foot, or only on the toes. The thumb is weakly developed and is usually lacking on the hind legs. All the toes have claws, which are either very blunt and adapted for digging or are curved and pointed and adapted for attack and defence. The front teeth are very small, the canines large and powerful, the premolars are knife-edged and work like a pair of shears, whilst the molars have rough biting surfaces and are sometimes used for crushing bones. Those Carnivores which are purely flesh-eaters have relatively small molars, but in the omnivorous forms the molars are large and flat. Many Carnivores have skin glands near the anus, which produce a strong-smelling secretion. The nipples are on the belly or in the groin, but never on the breast, and as a rule there are several pairs. The period of gestation is usually short and the female normally rears more than one young. The Carnivores may be divided into two main groups, one containing the cats, hyaenas and civets, the other the dogs, martens, raccoons, pandas and bears.

The Cat Family (Felidae)

In the cats the tooth series is very short and there is only one small molar in the upper jaw. The jaws are therefore short and this accounts for the short muzzle and rounded head. The body is slender, the tail usually long, and the claws strongly curved, pointed and laterally compressed. The claws can be retracted very strongly, so that they do not rest on the ground when the animal is walking; they are only extended when actually in use as weapons of defence or attack. This is why the tracks of a cat lack an imprint of the claws. The worn claws are sharpened by rubbing off the outermost horny layers; cats can often be seen

Lion and Lioness

sharpening their claws on the bark of trees. They all have a fine sense of hearing and well-developed sight, but the sense of smell is not particularly good. They feed almost exclusively on live vertebrate animals, which they usually lie in wait for or creep up to before attacking. The family has three main genera: the true cats, of which there are many species, the lynxes and the cheetah. (Some authorities place the cheetah in a separate family.)

Lion, *Felis leo,* 78+37 ⊥ 41 in. Before the use of firearms, lions were found all over Africa and southern Asia, and in classical times there were lions in Greece. In Africa they are now found only south of the Sahara, although scarcely ever in the Congo, in the extreme south or along the coasts. In India there are a few in the area north-west of Bombay, where they are protected. Lions live in open savannah country, along the edges of deserts and sometimes up in the hills. They are reddish-brown to greyish-yellow in colour with a black tuft on the tail, and the male has a prominent mane round the neck. There are several races, amongst them the now extinct *Barbary Lion* from the Atlas Mountains, with the mane extending down the back and along the belly; the relatively small *Somali Lion* with a weakly developed mane, large ears and a long tail; the long-limbed *Masai Lion* of East Africa; and the big *Cape Lion,* now extinct, with a black mane which extended to the belly. Lions feed principally on zebras and antelopes, but will also attack giraffes and buffaloes. They usually lie in wait for their prey at drinking-places and spring on to it without warning. They drag the prey into shelter and usually eat the entrails first. The breeding season is not restricted to any particular part of the year. After a gestation period of about 3½ months the female gives birth to 1-6 cubs, most often 2-3. Sometimes the cubs are blind at birth, sometimes they can see; their coat is spotted, and in the females may remain so until they are fully grown. Lion cubs are weaned rather late and they remain with the family for a long time. During periods when they are not mating the males often gather together into groups.

Tiger, *Felis tigris,* 80+37 ⊥ 41 in. Found in Asia from the Eastern Caucasus to China, and from southern Siberia to southernmost India, Java and Sumatra, but not in Ceylon or Borneo.

Tigers are most frequent in jungle country along rivers and lakes, and in the north on the steppes. The coat is usually rust-yellow with black transverse stripes, the belly and the lips are white; the largest tigers are somewhat bigger than the largest lions. There are several races, for instance, the *Indian* or *Bengal Tiger,* a powerful animal with very distinct markings; the *Siberian Tiger,* the largest form of all, which is somewhat paler in colour than the others, has fewer stripes and a longer and thicker coat in winter; and the *Persian Tiger,* a smaller form with shaggy fur. Tigers usually hunt for their food in the period just before and after sunset, but sometimes also at other times. They live principally on antelope, deer, wild pigs and monkeys, but also take sheep, goats and cattle; old tigers

Tiger

Leopard or Panther. (right) *ordinary form;* (left) *Black Panther*

may become man-eaters, and when this happens they are more dangerous than a lion. A tiger will lie in wait and pounce on its prey just as a lion does, but it always hunts alone. In the tropical parts of their range tigers may breed at any time of the year, but the Siberian Tiger has its cubs during the spring. The gestation period is only 3½ months and the tigress usually has 2-4 young. When they are 6 weeks old the cubs accompany the mother from one hiding-place to the next, and at 6 months they begin to hunt on their own.

Leopard or **Panther,** *Felis pardus,* 59 + 37 ⊥ 23 in. Found both in Africa and Asia; in fact leopards live throughout the range of both the lion and the tiger, and in addition occur in Asia Minor, Syria and Ceylon, and in Africa they are also found in the tropical forest regions. The *Black Panther* is not a separate species, but only a colour variety which is quite common in Sumatra, Java, Malaya, Assam and Abyssinia; in some lights one can still see the spots on the coat of a black panther. There are several races of leopard; for instance, the *Savannah Leopard* with a yellowish leather-brown ground colour and closely-placed spots, the *Desert Leopard* with a pale sandy-yellow ground colour, and the *Mountain Leopard* which is dark leather-brown with a black line along the back and a thick coat. Leopards are at home in all kinds of habitat and are wonderful climbers and good swimmers; when hunted they will often take shelter up in a tree. They are very bloodthirsty and kill much more than they can eat. They feed principally on antelope, jackals, monkeys and birds, but will also take cattle, goats, sheep, asses, dogs and fowls, and even mice, lizards or carrion. They usually start by eating the entrails and limbs, and carefully hide the rest away. They often drag the prey up into a tree and fix it firmly in the fork of a branch. The leopard is more feared as a man-eater than either the lion or the tiger. Mating takes place in the early spring and after a gestation period of only 3 months the female gives birth to 2-5 cubs, in which the eyes first open at 10 days. A female with cubs will hide away in a cave or under the root of a tree, but she takes them out quite soon to hunt at night.

Snow Leopard or **Ounce**, *Felis uncia*, 45 + 36 ⊥ 23 in. Found in Central Asia, particularly in the mountains regions of Tibet and the Himalayas. During the summer, snow leopards live very high up, but in winter they come down to altitudes of 6,000-9,000 ft. The coat is long and thick, mainly pale grey with a yellowish tinge but white on the belly, and marked all over with black spots. The cubs have broad, black longitudinal stripes on the hind part of the body, instead of spots. Snow leopards feed on wild sheep and goats, marmots, and birds, and will also attack small domestic animals. They are comparatively peaceful and there is no record of them attacking man, but they will immediately defend themselves if they or the cubs are threatened. They appear to be less nocturnal in their habits than the other large cats.

Puma, *Felis concolor*, 55 + 29 ⊥ 29 in., also known as the cougar or mountain lion. Pumas occur from British Columbia in Canada, along the Rocky Mountains, through Central America, in the Andes, and south to Patagonia; in South America they are found from the west right across to the east coast. They are mainly dark reddish-grey, being darkest on the back and lightest on the belly; the throat and inside of the ears are white, whilst the outside of the ears, a spot on the upper lip and the tip of the tail are black. The cubs have dark spots on the coat. Pumas live mainly on the fringes of large forests. They often spend the night in a tree, where they will also go if hunted. When chasing monkeys in the tree-tops they often jump from tree to tree. They also live on the plains among tall grasses. Most of their hunting is done at night, but they can sometimes be seen by day as well. The largest

Snow Leopard or Ounce

Puma

mating season is in March and after 3 months' gestation the female bears 1-3 cubs. The cubs start to lose the dark spots on the coat at about 3 months, and after their first moult in autumn they resemble the adults.

Clouded Leopard, *Felis nebulosa,* 39 + 33 $\perp$ 17½ in. Found in the East Indies and south-east Asia. Clouded leopards have rather short legs and a strikingly long tail, which is proportionately longer than in any other Carnivore. The coat is ash-grey or brownish-grey, the main part of the body being marked by large dark-edged greyish areas enclosing small dark spots. Clouded leopards are excellent tree-climbers, and feed mainly on small mammals and birds, but may also hunt on the ground for sheep, goats, pigs and poultry.

The Marbled Cat, 21 + 22 in., has a similar distribution to the clouded leopard, and is also an arboreal animal. It is not much larger than a domestic cat. The brown coat is marked with black stripes and spots on the back; there are also black transverse stripes on the sides which may be broken up into spots.

animals which they hunt are deer, sheep, calves and foals, but they will also take any smaller animals that they come across. When a puma has made a kill, it first of all opens up the neck and drinks the blood. If the prey is large it will then eat some of the front quarters and hide the rest under leaves and grass. Pumas are great wanderers and remain for only a short time in the same hunting area. In South America the

Jaguar, *Felis onca,* 70 + 25 $\perp$ 33½ in. Jaguars are distributed from the southern parts of the United States, through Mexico and Central America and south to northern Argentina. They are larger than the leopard, with a more powerful head and a shorter tail, but the colour pattern on the body is rather similar, except that in the jaguar there is often a spot inside each dark ring. Jaguars are seldom found in tropical forests, but are very frequent on the edges of forests, or along rivers and in marshy areas with tall vegetation, where they hunt pecca-

Clouded Leopard

ries, agoutis, alligators and terrapins; they also take horses, cattle, pigs and dogs. Unlike the puma, jaguars do not move about very much, and only change their hunting-grounds if persecuted, or if there is no more prey, or if there are floods. The male and female only live together for 4-5 weeks during the autumn. The gestation period is 100 days and there are two, or rarely, three cubs in each litter; the coat of the cubs is at first very pale in colour, and longhaired.

Jaguar

123

Ocelot

middle. Ocelots are good climbers and they hunt for monkeys and birds up in the trees. They live in the depths of the forest and in mountain country, usually remaining hidden in hollow trees or in thickets during the day and only beginning to hunt during the evening. The pairs remain together throughout the year and mating takes place in October-January; as a rule there are two cubs in a litter. Ocelots are much hunted for their beautiful pelts, which are highly valued in the fur trade.

Wild Cat, *Felis sylvestris,* 25+12 ⊥ 16½ in. Found in Scotland, central and southern Europe and Turkey, and eastwards to Turkestan and central Siberia. The upper parts are grey or grey-brown; the inner sides of the back legs are rust-coloured, and the whole coat is marked with dark bands and spots. The tail is pointed in the young, but in the adults it has the same thickness all along its length and is marked with black rings. Wild cats are found mostly in mountainous regions, where they use deserted fox earths, badger setts or hollow trees as places to shelter in. As night approaches they start to hunt for hares, rabbits or mice, small carnivores or birds; they also catch fish. Mating takes place in February, and after 9 weeks' gestation the females gives birth to 2-6 blind kittens.

Ocelot, *Felis pardalis,* 39+16 ⊥ 19 in. Found from Texas and Mexico through Central America and down to southern Brazil and northern Argentina. The ground colour is brownish-grey or smoky grey on the upper parts, yellowish-white on the belly; the patterning, which may vary considerably, consists of longitudinal black bands on the head and neck, and longish black patches on the body which have rust-coloured areas in the

Caffer Cat, *Felis ocreata,* 21+11 in. This species represents the wild stock from which our domestic cats have been derived; several races of it occur in Africa, and also in Syria and Arabia. The coat is yellowish or ash-grey with a rust-red sheen. The tail has dark rings and tapers towards the black tip.

Cats. (top, left) *Wild Cat;* (right, from above) *Caffer Cat, Black and White Cat, Tabby Cat, Tortoiseshell Cat, Siamese Cat, Red Persian Cat, Blue Persian Cat*

The **Domestic Cat,** *Felis catus,* is descended from the Caffer cat, which was regarded as a sacred animal in Ancient Egypt. Cats appear to have been domesticated for about 4,000 years; skulls from this period show that even then cats had the foreshortened face which distinguishes them from their wild ancestors. The spread of domestic cats to other countries took place relatively late. They were not commonly mentioned by Greek and Roman authors before the first century A.D., and began to be distributed through Europe only between then and 1000 A.D. In the 14th century they were still, however, quite rare in Europe and it was not until the 1700's that they became common as mouse-hunters in houses and farmyards. The fact that cats were originally sacred animals may have had something to do with the slow rate at which they spread throughout the civilized world. Domestic cats vary considerably in colour; they may be coal-black, snow-white, mouse-grey, fox-red or almost any other colour, and in addition, of course, many have dark transverse stripes. Many cats have two colours on the coat, but those with three colours are very rare and nearly always females.

In the **Siamese Cat** the slender, short-haired body

Serval

is yellowish-brown in colour, the legs, tail, face and ears are coffee-brown and the eyes sapphire blue.

Persian Cats came originally from Asia Minor. They have long, soft, silky fur, which may be white, yellow, red or blue-grey.

The *Manx Cat* from the Isle of Man has a stumpy tail or none at all, and although its legs are relatively long it is a good climber. There are also tailless cats in Java, Borneo and Japan.

Domestic cats usually have two breeding periods during the year, in early March and early June. The gestation period is 56 days and the litter contains 5-6 blind kittens, whose eyes open after about 9 days. They can mate with the wild cat. Cats are well known for their mouse-catching activities, and sometimes they take rats; they also catch and eat lizards, frogs, cockchafers, grasshoppers and small birds.

Serval, *Felis serval,* 30+15 ⊥ 10 in.

Found almost everywhere in the bush country of Africa, but not in Egypt. The coat is long, thick and coarse, and is rusty grey-brown in colour with large black spots; there are also completely black varieties. Servals hunt young antelopes, lambs, hares and small rodents as well as birds, and sometimes break in and do severe damage among stocks of domestic animals. They become active at nightfall, spending the day in rock crevices or among thick undergrowth, where they lie up and sleep. In East Africa servals mate in February-April and have 2-5 kittens in a litter.

The *Servaline Cat* is a separate species, similar in appearance to the serval but with more numerous and smaller spots on the coat. It also lives in Africa, but in the areas of tropical forest, where the serval does not occur.

The *African Tiger-cat* is smaller and has darker ears. The coat is brownish or greyish, and the distribution of spots is very variable; some have no spots. It lives in the tropical forests of central

Africa reaching as far west as Sierra Leone.

The *Lynxes* are distinguished from the true cats by their short tail, long legs, side-whiskers and the hairy tufts at the tips of the ears. They occur in Europe, Asia, Africa and North America.

The *Caracal Lynx* has strikingly long ear tufts and a greyish-yellow coat without spots. It is found in the deserts and open country of Africa and of western Asia as far east as India, but is absent from Ceylon. Caracals kill small antelopes, deer and birds. When attacking domestic fowls, a caracal will jump right into the middle of a flock and knock out several of the birds before they have a chance to fly away.

European Lynx, *Lynx lynx,* 33+5 ⊥ 28 in. Found in the mountain forests of northern Norway, Sweden and Finland, in eastern Europe and western Asia, but not in Siberia. The thick and rather long coat may be either yellowish, reddish-yellow, reddish-brown or greyish, marked with narrow flecks which vary considerably in number and position; in winter the coat is greyish-white with indistinct spots. The ears are erect, with long black tufts at the tips, and the stumpy tail has a black tip. Lynxes are nocturnal animals, hiding by day in thick undergrowth or cliff caves, and emerging at night to hunt hares or game; they also take deer, sheep and goats as well as foxes, squirrels and mice. After a gestation period of 10 weeks the female gives birth in May to 2-3 young, exceptionally 4; the eyes of the young open when they are 16 days old. *The Spanish Lynx* is a slightly

smaller animal, with more spots on the coat; it is found in Spain and Portugal.

The *Canadian Lynx* is similar to the European lynx, but its coat is longer and thicker, greyer in colour and with less prominent spots. It is found in the forests of Canada southwards to the Great Lakes and westwards to the Rocky Mountains.

European Lynx

Cheetah or Hunting Leopard

The *Cheetah* form the third genus in the cat family. They have long slender legs and a very small head. It is often said that the claws cannot be retracted but this is not strictly true; they can be retracted, but here the skin lobes which hide the retracted claws in other cats are lacking. They live in open country over almost the whole of Africa, apart from the tropical forest areas, and in Asia from Syria eastwards to northern India. The coat is usually pale yellow with a variable number of almost circular black or brown spots. The fur is not spotted in the young. As might be expected in an animal with such an enormous geographical range there are several races. One of these races, from Rhodesia, has traces of a mane on the back of the neck. The Indian form is smaller and its coat has fewer spots.

Cheetah or **Hunting Leopard,** *Acinonyx jubatus*, 53+29 ⊥ 31 in. Cheetahs prey on the smaller antelopes and deer and occasionally take sheep and goats.

Over short distances they are the fastest of all mammals, a speed of 70 miles per hour having been measured. Over long distances, however, they lack the endurance of the horse and dog. When a cheetah sights a herd of grazing antelope or deer, it creeps up to them, facing the wind, and then rushes forward, strikes down the prey with a fore-paw and bites it in the throat. From the earliest times cheetahs have been trained for hunting in North Africa, Abyssinia and India. Indian princes still use them as hunting animals and pay large sums to have them trained. Surprisingly enough it is the old animals which are caught for training, and not the young, which do not master the correct technique of hunting. A hunting cheetah will stop chasing the prey if it has not caught it after about 500 yards. The voice is very like that of a domestic cat, but deeper. When at rest a cheetah purrs almost continously, but if frightened or annoyed it has been observed to bare its teeth and produce a hissing sound.

Hyenas. (from above) *Spotted Hyena, Striped Hyena, Brown Hyena*

The Hyena Family (Hyaenidae)

Hyenas have the front legs considerably longer than the hind legs, so that the back slopes down towards the tail; the head is large and broad, the muzzle long and dog-like, the tail short and bushy and the fur long and coarse. Hyenas live in open country in Africa, western Asia and the western parts of southern Asia. They remain hidden during the day, but are active throughout the night and often enter villages in search of sheep, goats and cattle. Their main food, however, is carrion, which they can scent at great distances; they eat both flesh and entrails, and with their powerful jaw muscles and strong molar teeth they are able to crush hard bones. In many places hyenas and vultures form a combined sanitary squad, with the hyenas cleaning up at night and the vultures taking the day shift. Hyenas have a powerful voice, which sounds rather like loud laughter.

Striped Hyena, *Hyaena hyaena,* 39+15 ⊥ 29½ in. The fur is coarse and rather long, and there is a mane on the neck and the back. The colouring is greyish-yellow with dark cross-stripes and the bushy tail is either striped or self-coloured. Striped hyenas occur in North Africa with a southern limit in Senegal and Nigeria, in a great part of East Africa and in western Asia from the Mediterranean to Bengal. They go about alone or in pairs and feed almost exclusively on carrion.

Brown Hyena, *Hyaena brunnea,* 37+15 ⊥ 27 in. The dark brown, long-haired coat hangs down over the back like a cloak; there are transverse stripes on the legs. In habits brown hyenas resemble the striped hyena, but they are now found only in South Africa and Rhodesia.

Spotted Hyena, *Hyaena crocuta,* 50+14 ⊥ 31 in. This is the largest of the hyenas. The ears are short and rounded, there is no mane on the back and the tail is short-haired. Spotted hyenas occur in southern and central Africa as

far north as Abyssinia, where they are found up to heights of 12,000 ft. They hunt in packs, often doing great damage to flocks of sheep and goats, and sometimes also attacking cattle and horses; only rarely do they attack man. Like the other hyenas, they feed principally on carrion. The gestation period is about 3 months and the female lies up in a hole in the ground to give birth to one or two dark-brown young, in which the eyes are open at birth.

Aardwolf, *Proteles cristata,* 32+12 ⊥ 20 in. This is a peculiar, ungainly animal, looking rather like a small striped hyena. They have long front legs, a sloping back, a long mane on the the back and a bushy tail. The fur is pale yellow with black cross-stripes, and the outer half of the tail is black. They occur in South and East Africa. Aardwolfs are nocturnal and spend the day in an underground lair. Their teeth are small and simple in form with big gaps between them; they feed mainly on termites and other insects.

Aardwolf

The Civet Family (Viverridae)

The viverrids have a pointed snout, and an elongated body with short legs and a long tail. A pair of large glands near the rump secrete a musky oil, which is probably of importance during the breeding season. This secretion, which is known as civet, is used in the East in the manufacture of scent. Viverrids occur only in Africa and southern Asia, apart from two species which are found in southwest Europe. Many of them are very bloodthirsty and may therefore do great damage; on the other hand they are useful as killers of mice and rats. They are not all so exclusively carnivorous as the cats, for many of them feed on fruits. The civets track down their food in the same way as dogs, but creep up slowly and do not chase after the prey.

Large Indian Civet, *Viverra zibetha*, $31 + 17 \perp 15$ in. Found in southern Asia from India and Burma to southern China. The coat is dark yellow-brown with black stripes and reddish spots, and the tail is black with white rings. Civets usually go about alone and lie up during the day among tall grasses or in thick undergrowth; by night they hunt for frogs, crustaceans, birds and small mammals and now and then enter villages to attack poultry. The female gives birth to 3-4 young in May-June.

Large Indian Civet

Feline Genet, *Genetta felina*, $21 + 17 \perp 6$ in. Found in the tropical forests of West and Central Africa. The body is unusually long, the legs are short and there is a short mane along the back. Genets hunt and kill more or less anything they can overpower.

The *European Genet* is similar to its African relative, but it has a pale-grey coat with long rows of black spots. It occurs in southern France, Spain, Portugal and North Africa.

Feline Genet

Genets are kept as pets in parts of France and North Africa, primarily to act as rat-catchers.

Ichneumon or **Egyptian Mongoose**, *Herpestes ichneumon*, $25 + 17\frac{1}{2} \perp 7\frac{3}{4}$ in. Found in northern Africa, Sudan and Congo and also in Spain and Portugal. The legs are very short and the tail has long hairs at the base. The fur consists of rust-yellow wool hairs and long black and yellow guard hairs. Ichneumons usually live near to water and feed on small mammals, birds, snakes, lizards and insects, and are much feared as poultry thieves. The ancient Egyptians regarded them as sacred and even used to embalm them; nowadays they are tamed and used for rat-catching.

The Common Indian Mongoose is smaller than the ichneumon and has light silver-grey fur. It occurs in India

Ichneumon or Egyptian Mongoose

and Ceylon and has similar habits to the last species. Indian mongooses are easy to tame and often live in and around houses, where they keep down rats and snakes; but they may also kill poultry. Kipling's Rikki-tikki-tavi was a mongoose renowned for his victorious contests with cobras. Mongooses are not immune to the venom of cobras, although its effect on them is probably slower than on other animals.

Crab-eating Mongoose, *Herpestes urva*, $22 + 14$ in. The fur is dark brown marked with a whitish band on the neck. Crab-eating mongooses live along the rivers from Nepal to southern China and southwards to Burma. They have webbed toes and are more aquatic than any of the other viverrids. They feed mainly on frogs, toads, fish and freshwater crabs.

Crab-eating Mongoose

The Marten Family (Mustelidae)

The animals included in this family are smallish Carnivores with a long body, short legs and powerful paws armed with pointed claws, which cannot be retracted. Most of them climb well but they are equally at home on the ground. The family can be divided into three groups—the martens, the badgers and the otters.

The Martens

These animals walk on the toes and have strongly curved, short, sharp claws. Marten are primarily carnivorous, and many of them are highly prized as producers of top quality furs. They have long bodies, short legs and five toes on each foot. The blind and helpless young are born in holes in the ground or in hollow trees, and are carefully looked after by the mother.

Pine-marten

Pine-Marten, *Martes martes,* 19+9½ in. The fur is dark brown with a characteristic yellow spot on the throat, which ends in a single point. Pine-martens have a wide distribution in Europe and Asia, extending from Ireland in the west to the Himalayas and Manchuria in the east. They live in extensive forest areas and spend the day in hollow trees or in large deserted birds' nests, and only start to hunt at dusk. Their main prey is the squirrel, which they kill either in its "drey" or chase through the tree-tops, but they will in fact hunt any bird or mammal that they can overpower, and will also eat birds' eggs, frogs and insects, as well as plums and other fruits. The gestation period is about 9 weeks and the 2-4 blind young are born in March-April in a warm hole in a tree, in the deserted nest of a bird of prey or under a pile of branches. Only the female takes care of the young, which remain with her until the autumn.

Beech-Marten, *Martes foina,* 19+9½ in. In this species the fur is also dark brown, but the patch on the throat is white and it divides into two. Beech-martens occur over most of continental Europe, but are absent from England, Sweden and Norway. They can be found in the woods, but are more often seen in lofts, stables and other outhouses, where they shelter in the hay during the day; they may also go right into towns. They feed at night on small mammals, birds and juicy fruits and are very fond of eggs. Sometimes they break into dove-cots and hen-houses where they kill many more birds than they can eat, perhaps because they become excited by the frightened birds which are flapping around them; in such cases they will often drink some of the blood. Beech-martens are valuable as killers of rats and mice. The 3-5 blind young are born at the end of April or the beginning of May.

Sable, *Martes zibellina,* 19½ + 5½ in. The fur is dark brown or sometimes blue-black with white tips to the hairs, but the coloration varies considerably between the races. The main centre of distribution of the sable is the conifer forest area of central Siberia. Sables are good climbers, both on cliffs and in trees, and usually hunt for food early

Sable

in the morning. They eat squirrels and other small mammals, birds and eggs, as well as blueberries, rowanberries and cone-seeds; they also raid the honey stores of wild bees and sometimes catch fish by scooping them out of the water with the front paws. They usually have 3-5 young in May. From time immemorial sable has been among the most coveted of furs and large numbers of these beautiful animals are still caught for this purpose in traps and snares.

Beech-marten

Stoat, *Mustela erminea,* 11 + 4 in. In most Carnivores the male is larger than the female and this is especially marked in the stoat and weasel. Stoats have two moults in the year, one in spring which gives the brownish summer coat and one in autumn giving the white winter coat; the latter is pure white in northern and central Europe, but in southern England and further south it is brown, sometimes with whitish patches. The

Stoat

tail retains its black tip throughout the year. Stoats live in open country, especially in places where there are stone walls and undergrowth. Their most important food consists of mice, rats, water voles and moles, which they hunt both above and below ground; they also take reptiles, birds and eggs, and will jump on to the backs of hares and kill them by biting them in the neck and sucking the blood. When frightened they give off a strong-smelling secretion from a pair of anal glands. The gestation period is about 6 weeks and the 4-7 blind young are born in May-June, usually in a sheltered nest in a stone wall or under a tree stump. Although beneficial to man as mouse-hunters, many stoats are killed every year for their fur; the ermine of regal and other robes is stoat fur. They are found over almost the whole of Europe, and far into Asia. There is a closely related species in Ireland, which has a smaller body; its fur does not turn white in winter.

Weasel, *Mustela nivalis,* 8+2 in. This is a smaller animal than the stoat, the short tail lacks the black tip and the fur has shorter hairs. In Britain and in most of the rest of Europe, weasels are usually brown both in summer and winter, but in northern and central Scandinavia and in the Alps the coat is completely white in winter. Weasels are most abundant in open country, but they are not often seen because they are usually out and about only at night, and they do most of their hunting in mouse burrows. In fact they feed almost exclusively on mice, although in summer they also take eggs and small birds and will even attack hares. They have a warm nest in the ground, under the root of a tree or in a stone wall, in which the 3-6 blind young are born; they do not seem to breed at any particular time of the year. Weasels are widely distributed in Europe and western Asia.

Polecat, *Mustela putorius,* 17+7 in. in the male, smaller in the female. The coat consists of long brown guard hairs which scarcely conceal the yellowish

Weasel

Ferret (back), *and Polecat*

them killing and eating the rabbits down in the burrows. They are found domesticated over almost the whole of Europe.

Old World Mink, *Mustela lutreola,* 15+5½ in. A close relative of the polecat, with dark-brown fur and a pale spot on the throat. Mink are found in northern and central Europe and eastwards to northern Asia, but not wild in Britain. They swim and dive well. In appearance and habits they closely resemble the American mink, which is rather larger and has finer fur.

The *American Mink* lives near fresh water, where it feeds on fish, frogs, crayfish, snails and pond-mussels. When frightened it produces an evil-smelling secretion from the anal glands. The female gives birth to 5 or 6 young at the end of April. American mink occur wild from Hudson Bay and northern Alaska southwards to California and the Bay of Mexico. In many countries they are farmed for their fur. Sometimes they escape from these mink farms, and there are thousands of these feral mink in Sweden where they have become a pest.

under-fur of wool hairs. Polecats live in fields, meadows, marshland and along the banks of lakes and rivers. They feed mostly on mice, rats, moles and shrews, but also take many frogs, toads, grass snakes, vipers, fish, birds, insects, snails and worms, and in addition berries and other fruits; exceptionally they may do damage by taking poultry and eggs, but this is compensated for by the good they do in keeping down rats and mice. The gestation period is 8-9 weeks and the 3-7 young are born in May in a hollow tree, in a disused fox-earth, in a hole under a tree or in a stone wall. They are found over most of Europe, including Britain, but not in Ireland.

Ferret, *Mustela putorius furo,* 16+6 in. This is an albino domesticated form of the polecat. The fur is usually white or yellowish, but polecat-coloured ferrets are also found, which have been produced by matings between ferrets and polecats. Ferrets have long been used for rabbit-hunting; for this task they are often fitted with a muzzle, to prevent

Old World Mink

Libyan Striped-Weasel

areas of Norway, northern Sweden, Finland, Russia, northern Asia and North America. (The New World form is sometimes regarded as a separate species). They hunt mainly at night and feed on lemmings and other small rodents, but will also eat eggs and birds, and sometimes berries. They are particularly feared because they can kill reindeer. Wolverines have the reputaion of being insatiable, bloodthirsty predators, but it is probable that this view is somewhat exaggerated. The gestation period is about 9 weeks and the 3-4 small greyish young are born in March-April in a hollow tree, in a cliff crevice or in a snowdrift. For over 100 years attempts have been made in Norway to exterminate the wolverine and premiums have been paid for each one destroyed.

Ratel or **Honey-Badger,** *Mellivora ratel,* 33+6 in. This is a short-legged animal, which walks on the soles of its feet; it was once considered to be related to the badgers. Ratels are found over almost

Libyan Striped-Weasel, *Ictonyx libyca,* 13½ + 9¾ in. Found in north-east Africa. Striped-weasels go out hunting at night for anything they can kill, particularly small mammals and birds.

The Cape Striped-Weasel is black and white like the Libyan striped-weasel, but has a more bushy tail. It occurs in southern and eastern Africa, and it also produces a stinking secretion. There are several other striped-weasels in Africa.

Wolverine or **Glutton,** *Gulo gulo,* 31+6 ⊥ 17 in. The fur is usually somewhat darker in winter than in summer and it also becomes dark with age. Wolverines are found in the mountain and forest

Wolverine or Glutton

Ratel or Honey-Badger

the whole of Africa south of the Sahara and also in Asia from Arabia to India. They feed on small mammals, birds, reptiles, roots and fruits, and particularly on honey.

The Badgers

These walk on the soles of the feet and have powerful digging claws.

Badger, *Meles meles,* 29½ + 6 ⊥ 11¾ in. Found

Badger

over the greater part of Europe and into Asia. The fur is long, coarse and greyish; the head is white with a broad black stripe on each side, running through and behind the eyes. Badgers live mainly in small woods where they are not far from fields and meadows. As a rule they dig a rather large den, known as a sett. There is usually much more upturned soil at the mouth of a sett than there is outside a fox-earth; also a badger sett lacks the rank smell of a fox's home. Towards evening badgers leave the sett in search of food. In some woods one can find large patches of ground where they have moved aside moss and leaves in their search for earthworms, larvae, frogs, insects and mice, which form their main diet. They also eat oats, mast, berries, toadstools and roots as well as carrion. During the winter, badgers sleep for several days at a time in their warm setts, which are lined with moss, dry grass and leaves, but this is not a true hibernation. At the beginning of March the female gives birth to 2-4 very small young in the winter quarters; the young are blind for the first month and usually spend a couple of month in the sett. Badger pelts are too coarse to be of value as a fur, but the hairs are used for shaving-brushes and painting-brushes.

The *North American Skunk* has a long black coat with longitudinal white stripes. If surprised or attacked it brings the tail up over the back, turns its back to the attacker and squirts an evil-smelling yellow oily liquid from the stink glands for a distance of several yards.

The Otters

Otters walk on the soles of the feet and their toes are webbed. They swim by the bending of the body and tail, using the rather short legs to steer with.

Common Otter, *Lutra lutra,* 31 + 17 in. The fur is shiny and dark brown, but is usually bleached by the sun in summer. Otters are nocturnal animals and lead a quiet sheltered existence. They feed mainly on fish, but also eat crayfish, frogs and water voles and sometimes young birds, or adult birds which are sick. They take the larger fish to fixed eating-places on land and will often eat only the head and part of the back. In bad winters when it is difficult to find open waters, otters wander far and wide, but in general they live in rushy lakes, along rivers and streams, in fish ponds and also in harbours and bays. They have their lair, which is known as a holt, under tree roots or in soft earth and there is nearly always a tunnel running down and opening under the water. The nest part of the holt is dry and lined with moss and dried grass and there the 2-3 small young are born with mouse-grey woolly coats. The young do not go into the water for two months, by which time they have moulted, and they usually remain with the mother for about a year. Otters are found over most of Europe, including Britain, in North Africa and in parts of Asia.

Common Otter

The *Sea Otter* is the only Carnivore that lives exclusively in and near the sea; the back legs look rather like those of the seals. They occur in the Bering Sea and off California, and feed on sea-urchins and crustaceans. The soft fur is known in the trade as Kamschatka Beaver. They were once much hunted, but are now protected.

The Dog Family (Canidae)

Dogs have long legs and they walk on the toes of the rather small feet; they have only four toes on each hind foot. The head has a long jaw and the tail is of moderate size. Unlike many other Carnivores dogs cannot climb, but they are excellent runners. Their diet is mixed, but consists principally of flesh, often in the form of carrion. They are found in most parts of the world and are very prolific. The family includes the true dogs and the hunting-dogs.

The True Dogs

This group contains the vast majority of species, all of which have five toes on each of the front legs.

Wolf, *Canis lupus,* 49 + 15 ⊥ 33 in. The coat is greyish, with coarse hairs; in summer it has a yellow or red sheen and in winter it is a little paler in colour. The wolf is rather like an **Alsatian dog,** but may be distinguished by the straight hanging tail and the oblique position of the eye pupil. Wolves feed on mammals and birds and also take carrion and juicy fruits. When there is a shortage of food in winter they kill horses, cattle and deer, but seldom attack man. They keep mainly to the forests, and in winter they may form into large packs which wander far and wide. Wolves are distributed right across the northern hemisphere, but have been exterminated in Britain, Denmark, Holland and Switzerland; there are, of course, several races within this enormous area of distribution; in Greenland and northern Siberia, for instance, the wolves (polar wolves) are white, and in Florida and elsewhere there are black wolves.

The *Coyote* of North America is smaller than the wolf. It feeds mainly on carrion but also catches hares, rabbits, sheep, goats and birds.

Jackal, *Canis aureus,* 31 + 12 ⊥ 19 in. The coat is thick, coarse and yellowish-grey in colour, and the tail is bushy. Jackals live in the steppe regions of North Africa, in western Asia eastwards to India and in southern Russia and Greece, and hunt by night in noisy packs. They break into henyards and also kill lambs and goats, and may visit plantations.

Domestic Dog, *Canis familiaris.* Dogs have been domesticated for some 8,000 to 9,000 years. They have probably been derived mainly from wolf stock, although jackals may also have contributed something to the present race. In domestic dogs the skull and snout are foreshortened, the back is proportionately short and straight and the tail is often turned up at the tip. Mating usually occurs in February and August. After a gestation period of 63 days the bitch gives birth to 4-6 puppies which are blind for the first 10-12 days. Domestic dogs of all sizes and shapes have been developed by man, ranging from the mastiff, weight 180 lb. or so, to the toy dogs which may weigh only 1 or 2 lb. The dingo and the pariah dog do not fit into the general arrangement of domestic dogs.

(from above) *Jackal; Striped Jackal and Wolf; Dingo, Alsatian and Borzoi; Boxer and Spaniel; Dachshund, Great Dane and Dwarf Spitz*

Carnivores

Dingo, *Canis dingo*. At one time found throughout the whole of Australia, but now exterminated over large areas on account of the damage it does to sheep. It is thought that the dingo was first brought to Australia as a domestic animal, which later became wild.

The *Pariah Dog* of the Balkans, North Africa and southern Asia, is not truly domesticated but lives close to human settlements.

The *Pariah Dog* of the hundred different races of true domestic dogs recognised by breeders in Britain and North America. Several systems have been devised for grouping these animals, but here we may use the practical method of dividing them into two groups only–the sporting dogs and the non-sporting dogs. The former group includes Bloodhounds, Foxhounds, Greyhounds, **Dachshunds, Borzois, Spaniels, Boxers,** and the numerous different Terriers. The non-sporting dogs include the Bulldog, Mastiff, **Great Dane,** St. Bernard, Sheepdogs, Bull Terriers, **Spitzes** and Huskies.

Foxes. (from above) *Grey Fox, Common Fox, Blue Fox, Arctic Fox, Platinum Fox, Silver Fox*

Striped Jackal, *Canis adustus,* 34+13 in. Found in many parts of Africa. The fur is yellowish-grey and the tail usually has a white tip.

Grey Fox, *Urocyon cinereoargentatus,* 23+15 ⊥ 11 in. The coat is grey and shorthaired, with a rust-yellow belly. Grey foxes are found from northern America southwards to the northern parts of South America; in habits they much resemble the common fox.

Common Fox, *Vulpes vulpes,* 31+15 ⊥ 13 in. The fur is thick and may be very variable in colour. The legs are long and the body is slender, although in its thick winter coat a fox sometimes looks rather short in the leg. They live in woods, undergrowth and reed thickets, but are seldom seen for they are shy and wary and often hunt at night. When looking for food a fox keeps its snout close to the ground, but at other times the head is held erect and the tail downwards. Foxes do damage by eating poultry and hares, but also do much good by eating mice, which they will even dig up from underground. They also eat grasshoppers, dung-beetles and other insects, and in autumn berries and other fruits which are lying on the ground, or which they can reach. Sometimes they bury stores as reserves before the birth of the cubs or against winter shortages. The lair or earth usually has several access tunnels; the 5-8 young are born in it, usually in April. The vixen brings back food to the young and by June they are ready to wander about in search of insects or to accompany her on mouse-hunts. Foxes live all over Europe, including Britain, and extend far into Asia. In the United States and Canada the common fox is replaced by the *North American Red Fox*, a slightly larger animal with longer fur, which is much valued in the fur trade.

Arctic Fox, *Alopex lagopus,* 23+13 in. The thick woolly fur is grey-brown or grey with a greyish-yellow belly in winter. However there are some Arctic foxes–the so-called **Blue Foxes**–which are bluish-black or brownish-grey throughout the year. These are more valuable than the ordinary Arctic foxes and are reared extensively in fox farms; the thick winter fur fetches the highest prices. Arctic foxes are circumpolar in distribution in the northern hemisphere and are the only mammals which occur over almost the whole of Greenland; they are also found in Iceland and in northern Norway and Sweden, where they are however very rare. They feed on lemmings, mice, ptarmigan and other birds, but will also eat carrion, and on the coast they take stranded fish. The pale grey-brown cubs are usually born at the end of May, and there are most often 6-8 in a litter.

The **Silver Fox** is a colour variety of the North American red fox and is farmed for its attractive soft fur.

The **Platinum Fox** originated as a mutation from the silver fox and is extremely valuable as a fur animal.

Fennec Fox

Maned Wolf, *Chrysocyon jubatus,* 35+15 ⊥ 27 in. This is the largest member of the dog family found in South America. The body is proportionately very short and the legs surprisingly long. There is a mane on the back of the neck which can be erected. The fur is cinnamon-brown with the belly somewhat paler; there is a brownish-black area on the neck and front part of the back, and the tip of the tail is white. Maned wolves are found on the plains of southern Brazil, Paraguay and northern Argentina. They hunt for food at night, taking mice, fish, sugar-cane and fruit.

Fennec Fox, *Megalotis zerda,* 17+8 ⊥ 8 in. This is the smallest of the foxes but it has relatively enormous ears. The soft coat is sand-coloured, with the belly white; the tail is rust-yellow with a black tip. Fennec foxes are found in and around the Sahara, and in the Sinai Peninsula and Arabia. Their food consists of small mammals, birds, lizards, insects and fruit. They are nocturnal and spend the daytime down in a lair.

Raccoon-like Dog, *Nyctereutes procyonoides,* 23+5 ⊥ 8 in. Found in Japan, North China and northwards to Manchuria; they have been introduced as fur animals into Russia, whence they have spread to Finland and Sweden. The legs are short and slender, and the thick fur is grey-brown in colour with a dark area on the shoulders extending down over the front legs. Raccoon-like dogs remain hidden during the day and come out at night to hunt in groups for mice and fish; they also eat berries and other fruits.

Maned Wolf

Hunting Dogs

Cape Hunting-Dog, *Lycaon pictus,* 43+15 ⊥ 25 in. Recog-

Raccoon-like Dog

over the forehead. The tail is yellow at the root, black in the middle and white at the tip.

Cape hunting-dogs are found throughout the whole of the open savannah country which surrounds the rain forest regions of Africa. They are active by day and night and usually hunt in packs, often 30-40 together. As a rule they hunt mostly for antelope but may also attack sheep. When hunting antelope, which rarely run away in a straight line, they will hold back and reserve their strength so as to be able to attack when the prey has become exhausted; during these hunts they bark and howl, creating an impression of great activity. When they attack a flock of sheep they usually kill many more than they can eat; often they will only tear out the entrails. The female digs a hole in the ground where she gives birth to up to 10 pups, but she does not seem to trouble herself much with them and will leave them in the lurch as soon as there is any danger.

nized by the big broad ears, short muzzle and bushy tail. The coat is mottled black, white and yellow, and scarcely two skins show the same patterning; the muzzle area is black and there is a black stripe running backwards from the snout

Cape Hunting-Dog

(above) *Raccoon;* (below) *Ring-tailed Coati and White-nosed Coati*

The Raccoon Family (Procyonidae)

The raccoon family walk on the soles of the feet and have five toes on each foot. They have long tails and live mainly in trees.

Raccoon, *Procyon lotor,* 22+10 ⊥ 13 in. Found in the southern parts of North America. The coat is mainly grey-black or silver-grey, with a black and white face and a bushy tail with black rings. Raccoons feed on birds, fish, eggs, crustaceans, snails, bivalve molluscs and insects, and also eat fruits. They usually wash the food before eating it. In the northern part of their range raccoons hibernate in the coldest months. The female has 4-8 young, often in a hollow tree. Thousands of their skins are marketed every year.

Ring-tailed Coati, *Nasua nasua,* 24+17 ⊥ 11 in. Found in South America. The fur is reddish-brown on the back and yellowish on the belly. Coatis live communally in forest areas; they climb well and feed on plants and small animals. The female gives birth to 3-6 young in a hole in the ground.

Cacomistle

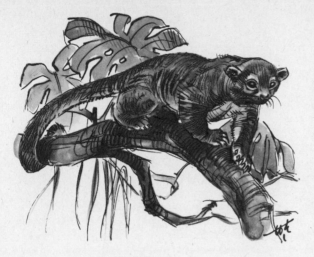

Kinkajou

The **White-nosed Coati,** *Nasua narica,* is a closely-related species, which occurs from southern Mexico to Peru.

Cacomistle, *Bassariscus astutus,* 20 + 14 in. A dark grey-brown animal with a long ringed tail, found in the southern United States, Mexico and Panama; it feeds on poultry, mice, rats, insects and fruit.

Kinkajou, *Potos flavus,* 17 + 18 ⊥ 6½ in. Found in the tropical forests of South and Central America. The head is thick and short, the body elongated, the legs short and the tail very long and prehensile. The fur is thick and yellowish-grey. Kinkajous are mainly nocturnal, feeding principally on plants, but also taking small mammals, birds, eggs, insects and larvae. They are very fond of honey.

Red or **Common Panda,** *Ailurus fulgens,* 24 + 18 ⊥ 13 in. The thick, soft, long-haired fur is shiny red on the upper side with a yellow sheen on the back; the belly and legs are silky black and the rust-red tail is marked with rather indistinct dark rings. The red panda is now usually placed in a special family–the Ailuridae–closely related to the raccoons. Pandas are found from Nepal over the eastern Himalayas to south-western China, at heights of 6,000-12,000 ft. Like martens, they live in tree-tops and seldom come down to the ground, where they can however move well by running and jumping. They feed mainly on fruits, buds and roots, but also take eggs and small birds. They sleep during the day and wake up as it gets dark. The two young are born in spring.

Red or Common Panda

The Bear Family (Ursidae)

The bears are large animals with proportionately low hind-quarters, powerful limbs with broad-soled paws, small round ears and a rudimentary tail; all the paws have five toes which are furnished with large strong claws. The teeth are adapted more for eating plant food, which has to be chewed and crushed, than flesh which only has to be cut into pieces. All the bears walk on the soles of the feet, which are naked, except in the Polar bear. The fur is particularly thick in winter and is usually very long-haired around the neck, on the belly and on the hind quarters; the very short tail is almost completely hidden in the fur. In summer bears eat so much that they accumulate a layer of fat, which often forms a kind of fatty hump in the shoulder region. Most bears sleep during the day in their shelter and move out at twilight. The bear family has representatives in Europe, Asia and America, but not in Africa or Australia; the majority of species are found in Asia, from the Arctic regions southwards to the Tropics.

Brown Bear, *Ursus arctos,* 82+3 ⊥ 47 in. The thick, brown coat may vary considerably in colour; in Norway the brown bears are reddish-brown with black feet, those in south-east Europe have a pale collar, and those in South Russia have a pale yellowish pelt. They live mainly in extensive areas of forests and in mountains, where they search for their food at night or from the afternoon right on till the following morning. Un-like most other Carnivores they feed mainly on berries, mast, toadstools, grasses, roots and honey, but will also take animals, from insects to mice, fish, birds, sheep, cattle and horses. In autumn they look for a shelter under a fallen tree or in a cave, where they gather together spruce twigs, moss and lichen to form a home for the winter. They do not, however, stay the whole winter in seclusion, and are often seen outside the den during the cold months. In fact bears do not undergo true hibernation, for although their activity is reduced, their body temperature does not go down, as it does in the true hibernating mammals such as bats and hedgehogs. In January the female has 2-3 blind young; they are only about the size of a rat, and weigh only 10 oz., which is only $1/300$ of the weight of the mother. For comparison a new-born calf weighs $1/30$ and a puppy $1/20$ of the mother's weight. Bear cubs are blind for the first 4-5 weeks and remain with their mother until the end of the summer, or even longer. In spite of their heavy build bears are good climbers, particularly when young. In Europe brown bears are still found in Norway, Sweden, Finland, Russia and in the Carpathians and Balkans, but have been completely exterminated in Germany, Denmark, Holland, Belgium, Switzerland and Britain. The **Isabelline Bear** is a race of the brown bear found in the Himalayas.

The Kodiak Bear, the largest form of brown bear, is found on Kodiak and neighbouring islands off Alaska.

The **Grizzly Bear,** *Ursus horribilis,* 98 in., weighs about 900 lb; it is found in western North America. Although here they are placed as a separate species,

Bears.
(top, right) *Brown Bear, typical form,* (left) *Isabelline form;* (centre) *Grizzly Bear;* (below,
left) *Black Bear,* (centre) *Cinnamon form of Black Bear,* (right) *Himalayan Black Bear*

Polar Bear

some authorities regard the grizzlies as races of the brown bear.

American Black Bear, *Ursus americanus,* 79 ⊥ 39 in. A glossy black animal, with yellow-brown around the tip of the snout, found in the forest areas of North America. Black bears are smaller and less aggressive than brown bears. Mating takes place in June-July and the cubs are born in the middle of January. The **Cinnamon Bear** is really only a brown variety of the black bear, and should not be regarded as a separate species.

Himalayan Black Bear, *Selenarctos thibetanus,* 74 ⊥ 32 in. Found in central Asia, from Persia and the Himalayas to China. They have a white or yellowish marking on the chest, but are otherwise coal-black except for the snout which is brownish. They are good climbers and are vegetarian, sometimes doing damage to plantations. They go into winter quarters in October, and remain there until the middle of March.

Polar Bear, *Thalarctos maritimus,* 98 + 7 ⊥ 55 in. A brown bear may weigh 500-750 lb. but a Polar bear may be up to 1,600 lb. The thick coat consists of short woolly hairs and long smooth guard hairs; the fur is shortest on the head, neck and back, but the hairs on the sides, belly and hind-quarters are often quite long. In winter the fur is almost pure white, but in summer usually yellowish-white. Polar bears live on and near the drift-ice and occur along the north and east coasts of Greenland, and in smaller numbers along the Arctic coasts of mainland America and Asia; occasionally they travel with the drift-ice to Iceland and the coast of northern

Norway. On the east coast of Greenland they drift southwards on the ice and then wander back north again along the shore. They swim well using the broad paws, and a thick layer of fat under the skin allows them to remain in the water for hours at a time. They may often be found swimming many miles from land. They are as good at diving as they are at swimming.

Polar bears feed mainly on seals, which they kill with a blow from the front paw; they approach such prey by swimming in among the ice-floes, or by creeping up over the solid ice, or sometimes they lie in wait at the breathing-holes which the seals keep open in the ice. They also eat fish, carrion, lemmings, bivalve molluscs and crustaceans, as well as some plant food, such as moss, grasses, bilberries and Arctic willow. In winter the pregnant female digs a hole in the snow with a very long entrance tunnel. In this shelter she gives birth in January-February to 2 blind cubs, which are no larger in size than a rat, and therefore relatively smaller than the new-born young of the brown bear. The females look after the cubs for about 6 months. Sometimes the males and non-breeding females also make a shelter to which they retire during the coldest parts of the winter.

Malay Bear (above); *Sloth Bear* (below)

Sloth Bear, *Melursus ursinus,* 70 $\perp$ 33 in. Found in mountain and forest country in India and Ceylon, including the foothills of the Himalayas. Sloth bears differ from the bears already described in having the lips almost completely naked. The claws are very long and sharp, and the fur, especially on the neck and sides, is black and shaggy. The front part of the head is greyish, and there is a halfmoon-shaped patch on the chest. Sloth bears sometimes eat birds and eggs, but they prefer juicy fruits, roots and honey, and they can tear open hard termite colonies with their powerful claws and lick up the termites. Sometimes they climb up palm trees and drink the contents of vessels hung there by the natives to collect sap for palm wine. The female gives birth to 2-3 young, which she carries about on her back when out searching for food.

151

Giant Panda

claws. The fur is black with the exception of the pale yellow or greyish forehead and the large yellowish or greyish spot on the throat. Malay bears are found in south-east Asia and the East Indies and are the best climbers of all the bears. They feed almost exclusively on plant food and insects, although now and again they may eat small mammals and birds. When hunting for insects they use the long narrow tongue which can be shot far out. They often climb up into coconut palms to eat the young shoots and may thus do much damage in coco plantations. It is peculiar that the new-born young are about the same size as those of the much larger brown bear.

Giant Panda, *Ailuropoda melanoleuca,* 65 in. A bear-like Carnivore, which is considered to be more closely related to the red panda and the raccoon family than to the true bears. The head is broad, with a short muzzle. Most of the body fur is white, but the ears, a ring around the eyes, the limbs and a broad band over the shoulders are all black. Giant Pandas live in the mountain and forest regions of eastern Tibet and south-western China. They feed mainly on plants, particularly on young bamboo shoots, but are known also to eat small mammals and birds. They are regarded as one of the rarest mammals in the world.

Malay Bear, *Helarctos malayanus,* 54 ⊥ 27 in. This is the smallest of all the bears. It has a short broad head, small ears and large paws with very long

Rodents

THIS order contains approximately a third of all the land mammals. Their most characteristic features are the strongly curved incisor teeth, of which there are one or two pairs in the upper jaw and one pair in the lower jaw. When used for gnawing, these incisor teeth become much worn, but they have open roots and keep growing throughout life. In most rodents there is enamel only on the front surfaces of the incisors, and as this enamel is harder than the dentine making up the main part of the teeth, the incisors become more worn on their rear surfaces; this accounts for the sharp chisel edges, which are so characteristic of the rodents. In the lagomorphs (hares and rabbits) the enamel extends right round the incisors, but even here these teeth become chisel-shaped, because the enamel on the front surface is harder and more resistant to wear. None of the rodents have canine teeth, and there is therefore a gap between the incisor and molar teeth. The molars have high crowns, and there are usually fewer of them than in most other mammals. The rodents are mostly vegetarian. On account of the enormous numbers of individuals, rodents play an important part in natural communities, for they provide the main supply of living food for many of the predatory mammals, birds and reptiles. On the other hand they breed fast and can therefore afford this tax on the population. Rodents have four toes on the front feet and five on the back feet. They are found in all parts of the world and almost everywhere that mammals are able to live: in the coldest as well as in the warmest regions, in deserts, in tropical forests, on the ground and in trees, in water and on land; they have not, however, become adapted for life in the air or out in the sea. There are two main suborders: the lagomorphs and the true rodents.

The Lagomorphs

The lagomorphs, or double-toothed rodents, have two pairs of incisor teeth in the upper jaw, the outer pair being small and placed behind the large inner pair. The upper lip is cleft, and the two lobes can be pushed to the side revealing an area of naked skin between them. There are two families: the pikas and the rabbits and hares. Nowadays most zoologists regard the lagomorphs as a separate order on their own, but still closely related to the true rodents.

Siberian Pika

on the upper side, whilst the belly and legs are ochre-yellow; some specimens are completely black. The ears are almost naked on the outer side. Siberian pikas are found in the mountain regions from Altai to Kamschatka and western China. Siberian pikas dig burrows and form colonies, often with more than 1,000 animals. At the beginning of summer they collect large stores of grass which they place in stacks 4-7 in. high and cover with large leaves. In certain areas with heavy falls of snow in winter the Mongolians drive their sheep out to eat this stored grass or else collect it for horse fodder, with the result that the pikas die of hunger.

Pikas

In contrast to the hares and rabbits the pikas have the front and back legs of about equal length; the ears are short and broad, and the large broad incisor teeth in the upper jaw have a deep longitudinal groove, so that they have two points. The thick close fur on the under side of the feet helps to prevent them sinking too deep into the snow. The tail is very short and hidden by the fur. Pikas live either in mountain areas or on the plains and have two main centres of distribution, one in Asia and one in North America.

The Dwarf Pika, which is not more than 5½ in. long. has soft grey-brown fur. It is found on the steppes of western Siberia and extends westwards to the Volga. It peeps like a small bird.

Siberian Pika, *Ochotona alpinus,* 10 in. The thick, short-haired fur is brownish

Rabbits and Hares

All the rabbits and hares are good runners with long back legs, long ears, large eyes, a deeply cleft upper lip and long whiskers. The fur consists of thick soft wool hairs and long guard hairs. Their sense of smell is well developed and their hearing is also good. As a rule they are mute but when frightened they give a loud wail. When alarmed they stamp with their back feet.

Mountain or **Blue Hare,** *Lepus timidus,* 23 + 2½ in. The tail and ears are a little smaller than the brown hare. Mountain hares moult both in spring and autumn, but not all of them become completely white in winter; some have a grey or blue-grey winter coat. They live mostly in woods or on the edges of forests, but may also be seen in open country. They feed on grass, corn, kale, berries, bark and shoots. There may be three litters in the year, in May, around the first of July and in the middle of August, each with 3-5 leverets. These are born with thick fur and open eyes, and can run around almost immediately. Hybrids occur in areas which have both mountain and brown hares. Mountain hares are found from Scotland eastwards over Scandinavia and northern Asia to Japan. There is a closely related form in Ireland, which does not usually turn white in winter; it is the only hare found in Ireland.

Mountain or Blue Hare (left, back), *in moult* (front); (right) *Arctic Hare*

Arctic Hare, *Lepus arcticus,* $27\frac{1}{2} + 3\frac{3}{4}$ in. These are larger in size than the mountain hare, especially the more northerly races. They are found on the tundras of North America, on the islands to the north of Canada and in Greenland. In the extreme north, on the islands north of Baffin Land and in north Greenland they are white the whole year round, but in the more southerly parts of their range they are grey on the back in summer; the tips of the ears are always black or dark grey. The food consists of Arctic willow and anything else they can get hold of in the way of green plants and berries. Mating starts at the beginning of May, when the large winter flocks break up into pairs. In June, when the short arctic summer is just starting, the females give birth to the only litter of the year, which usually has 5-6 grey leverets. Falcons, snowy owls and ravens prey on the Arctic hares from the sky, and wolves and foxes chase them over the snow.

Brown Hare, *Lepus europaeus,* $26 + 3 \perp$ $11\frac{1}{4}$ in. Found particularly in open fields, and also in meadows, dunes and small woods. The short front legs and the very long back legs make them walk

rather awkwardly, but they jump remarkably well and can run at a speed of 50-60 miles per hour. The feet have no pads, but the soles are covered with a thick layer of hairs. The tracks of a hare are easy to recognize in snow or soft ground, for the long spoors of the two back feet are alongside each other and in front of the smaller tracks of the forefeet. This is because the leap or bound of a hare is so powerful that the back feet land in front of the forefeet. When listening, hares sit on the hind legs with the long ears erect. They lie up for most of the day in among tall grass in a little hollow, known as a hare's "form", and usually go out to feed at dusk or during the night; they eat chiefly grass, but also nibble at clover, corn, turnips, cabbage and other greenstuff. In bad winters they may do damage by eating the bark, buds and shoots

Brown Hare

and most of Sweden. There are several other species of hares in Africa, Asia and America, all of which are similar in habits and appearance to the brown hare.

Wild Rabbit, *Oryctolagus cuniculus,* $16\frac{1}{2} + 2\frac{1}{4}$ in. Rabbits can be distinguished from hares by their shorter ears which do not have black tips, and, of course, the limbs are also shorter. They live in extensive burrows or warrens which have several emergency exits, and they feed mainly on grass, clover, corn, turnips, kale and bark. The 3-12 young are born in an underground nest and are naked and blind at birth. The eyes open when they are about a fortnight old and they suckle for 3-4 weeks. During the summer the female may have one litter after another, at intervals of 5-6 weeks. Rabbits came originally from south-west Europe, whence they have spread over most of Europe. At the present time the wild rabbit populations in Britain and Australia have been much reduced by the spread of the disease known as myxomatosis.

Red Rabbit, *Pronolagus crassicaudatus.* Found in large colonies in the mountains of South Africa. There are several other species of rabbits in Africa, and in Asia and America.

Domestic Rabbits have been derived from the Wild Rabbit, and there are now more than 50 different races. Among the giant-sized races is the **Flemish Giant,** which may weigh 12-14 lb. or even more; it is usually steel-grey in colour and long in the body, and it has long ears. In the **Lop Rabbit** the ears are very long and drooping and the fur is thick. The **Angora Rabbit** is medium-sized and usually white; the long hairs

of young trees. The mouth is small but as the upper lip is cleft they can work away freely with the incisor teeth. The females produce 3-4 litters in the year, each with 2-5 young. The new-born leverets have long thick fur and open eyes and can move around almost at once. During the greater part of the day they lie hidden, and the mother only comes to them when they need to be fed, and after about 3 weeks they can look after themselves. Hares are hunted by foxes, stoats, and dogs; they seldom fight back, but either lie low in the form or else bound off at speed. They are found over almost the whole of Europe, with the exception of Ireland, Norway

Rabbits. (from above) *Wild and Red Rabbits; Lop and Angora Rabbits; Ermine Rex and Flemish Giant; Blue Vienna and Dutch Rabbits*

are used as wool. The **Blue Vienna** is a fast-growing race with rather thin fur, still commonly kept on the Continent. The **Dutch Rabbit** is one of the commonest of the domestic rabbits; the fur is usually marked black and white, although other colours occur. The **Ermine Rex** is a small rabbit, nowadays mostly albino and therefore having reddish eyes.

The True Rodents

In the true rodents each jaw has only one pair of incisor teeth, and these have enamel only on their front surfaces. The majority of the rodent families are very similar to each other. Here we will describe representatives of some sixteen families, but there are about ten others.

Squirrels

Squirrels are adapted for life in the tree-tops. The long bushy tail is used as a rudder during their aerial jumps and the claws, which are more pointed and curved than in the digging rodents, are well adapted for gripping. There are about 200 species distributed in all parts of the world except Australia and Madagascar.

Red Squirrel (front); *black form* (back)

Red Squirrel, *Sciurus vulgaris,* 10+7½ in. Easily recognizable by the long bushy tail and the hairy ear tufts, which are long in winter but may be absent in summer. Squirrels are diurnal animals, which climb about very fast in trees by gripping the bark with their sharp curved claws. They also jump by pushing off with the powerful back legs and steering with the tail which is held out behind; in this way they may cover distances of several yards. When they sit still, the tail is usually held erect over the back. Squirrels feed chiefly on pine-cones which they fetch from the ends of the slender branches of conifers. Once a squirrel has bitten a cone loose, it climbs back to the trunk of the tree, sits down with the cone between the front paws, and gnaws off the scales to get at the seeds. When only the uppermost scales are left it lets the remainder of the cone fall to the ground. In a cone which has been gnawed by a squirrel the axis most often ends in a point at the bottom, for there are usually no seeds under the lowest scales, which are therefore pulled off quickly, so that the axis becomes frayed and pointed. Squirrels also eat hazel-nuts by gnawing a hole in the shell with the front teeth of the lower jaw. When they find more nuts than they can eat, they often bury them in the earth or hide them in a safe place. They make their round nests, which are known as "dreys", up among the branches, and line them with moss and hair. They sleep in the small nests or in hollow trees, but breed in the larger nests. The gestation period is about 35 days and the 3-7 young are born blind and almost naked in March-April; there may be one or two more litters later in the year. Besides nuts and cones they also eat mast, acorns and toadstools, and probably also take eggs and young birds. They are hunted by buzzards, goshawks, beech martens and pine martens. There are several races of the common squirrel distributed over the whole of Europe and through Siberia to northern China.

The *American Grey Squirrel* has been introduced into Britain and has now spread over much of the country. It is a larger animal than the Red Squirrel and far more destructive. Efforts are being made to control the spread of this immigrant.

Indian Giant Squirrel, *Ratufa indica,*
19 + 17 in. This is a large species about
the size of a marten, which lives almost
entirely in the trees, where it builds an
extensive nest of branches and leaves.
It is found in India.

Flying Squirrels

This family has a large number of species distributed throughout the northern
hemisphere, particularly in south Asia.
There is a flight membrane running
along each side of the body between the
front and back legs. When this flight
membrane is spread out it acts as a kind
of parachute which helps to slow down
the speed of descent when the animal
makes a long jump. Flying squirrels
feed at night on bark, buds, fruits and
insects.

*Indian Flying Squirrel, and North American
Flying Squirrel* (below)

Indian Giant Squirrel

Indian Flying Squirrel, *Petaurista phi-
lippensis,* 23 + 23 ⊥ 7½. Indian flying
squirrels live in tall trees in India and
Ceylon and can glide for distances up
to 180 ft.

The *European Flying Squirrel,* 6¼ +
3¾ in., is much smaller than a red
squirrel and has a thick, soft, silky coat.
It is found from western Finland
through north Russia into Siberia, mainly in birch woods.

North American Flying Squirrel, *Glau-
comys volans* and *G. sabrinus,* 5 + 3¾
in. Found in the forests of western
North America. The fur is greyish on
the back and almost silver-white on the
belly. They sleep soundly during the
day, often in clusters of 10-20 animals.
They can glide for up to 90 ft.

Marmots

This group includes the marmots, prairie dogs, sousliks and chipmunks. They all have cheek pouches and most of them dig well. They are found in south-east Europe, north and central Asia and North America.

The *Marmots* have a powerful body, a rather short tail, short ears and small eyes. There are a few species in central Europe, north Asia and North America.

Alpine Marmot, *Marmota marmota,* $20 + 4\frac{1}{2} \perp 7$ in. The coat is red-yellow with ash-grey on the top of the head and on the back; they have no thumb. At one time alpine marmots were widely distributed but are now found only here and there in the Alps and Carpathians. In summer they keep near to the tree limit and above, living either alone or in pairs in underground burrows which have several exits. In winter they move to lower altitudes and live in large groups in deeper burrows which they line with dry grass. They hibernate for more than half the year, and during this time their body temperature may drop below $41\,°F$. The female gives birth to 4-6 grey-blue young in June. The warning signal of a marmot is a piercing whistle.

Bobac Marmot, *Marmota bobak,* $32 + 4 \perp 7\frac{3}{4}$ in. Found on the steppes of South Russia and far into Asia. The fur is sandy or rusty yellow in colour and much resembles that of the Alpine marmot. Bobaks start to get ready for the winter sleep as early as June; in September they stop up the entrances to their burrows, but do not start to hiber-

Alpine Marmot

nate properly until December. The young are usually born in April-May.

The *Prairie Dogs* live in very large colonies in North America. They get their name partly from their habitat and partly from their characteristic dog-like bark. They have well-developed thumbs with powerful claws.

Black-tailed Prairie Dog, *Cynomys ludovicianus,* $13\frac{1}{2} + 2\frac{3}{4}$ in. Found on the prairies and in the Rocky Mountains from southern Canada to northern Mexico, the black-tailed prairie dog is one of the most abundant and striking wild animals of the prairie; they have been almost exterminated in the cultivated areas. The fur is red-brown on the back and greyish on the belly; the outer third of the tail is black. Each family has its own deep burrow with a small mound of earth on the surface. Although there is no sign of any division of labour such as occurs in true colonies of animals, the prairie dogs do have lookouts which give warning of danger. They feed on grass and roots. The female gives birth to 4-6 young in May.

The smaller *White-tailed Prairie Dog* is found in the southern parts of the Rocky Mountains.

The *Sousliks* form yet another group with several species. They are smaller and more slender than the prairie dogs and have small ears and large eyes. The cheek pouches are large and the tail is short and hairy.

Bobak Marmot

Common Souslik, *Citellus citellus,* 8½ +3 in. Found in Saxony and Czechoslovakia, in the north-eastern part of the Balkans and in South Russia. The fur is greyish-brown on the back and rust-yellow on the belly. Sousliks dig long burrows in the earth, and make their home more than 3 ft. below the surface. Sousliks feed on corn, clover, roots and berries as well as on insects, mice, small birds and eggs. They hibernate from November to April, and the female gives birth to 3-8 young in April-May.

Spotted Souslik, *Citellus suslica,* 8+2 in. Found in eastern Europe, where they do a lot of damage in maize plantations. The coat is grey or rust-brown with whitish-yellow spots on the back and sides. They hibernate from September to March and many are frozen to death during the winter.

Black-tailed Prairie Dog

Spotted Souslik (back), *and Common Souslik*

Leopard Souslik (back), *and Eastern Chipmunk*

Leopard Souslik, *Citellus tridecemlineatus,* $7\frac{1}{2}+3\frac{3}{4}$ in. Found throughout the whole of the prairie region east of the Rocky Mountains. Although they do much damage in the fields by eating planted seed, they also consume large numbers of grasshoppers and other harmful insects, and sometimes take rats and mice. The burrows are not particularly far beneath the surface, and although the animals remain in them from October to April they are scarcely hibernating the whole time, since they store large amounts of corn. The female gives birth to 7-10 young in May.

The *Chipmunks* have large cheek pouches which extend right back to the shoulders. They occur in North America and North Asia and they often have dark or light longitudinal stripes along the body.

Eastern Chipmunk, *Tamias striatus,* $6+3\frac{1}{2}$ in. Found in eastern North America. Chipmunks collect large stores of corn and they hibernate together in large clusters. When conditions are good the females may have two litters in the year.

Beavers

The family contains only one genus with two species, the European beaver and the North American beaver, which are very similar to each other in appearance and habits. They differ from the other rodents in having broad flat scaly tails, which are almost hairless.

European Beaver, *Castor fiber,* $33+15$ $\perp$ 11 in. This is the largest of all the Europeans rodents. The fur is long and thick and varies in colour from pale to dark brown, the back usually being the darkest part. There is a double claw on the second toe of the back foot which is evidently used for cleaning the fur. Beavers swim remarkably well with the back legs which have webbed toes, and use the flat tail as a rudder. They live in rivers and lakes where they build elaborate dams of branches and mud. One or more underground passages lead off from the "lodge" and open out under the water. The dams are built of branches and are made downstream from the lodge, so that the water rises around it to form a pool. They feed mainly on tree bark, and also on leaves, buds, shoots and roots. They are able to fell quite large trees by gnawing away at the trunks at ground level; for this they use the very powerful front teeth which have orange-yellow enamel. They store young branches on the bottom of lakes to act as a food reserve for the winter. Mating takes place from the middle of February until the end of March, and the female gives birth to 1-3 rat-sized young at the beginning of May. At one time beavers were widely distributed throughout Europe, but nowadays they are common only in southern Norway where there are several thousands. There are also populations surviving on the Elbe, in the Rhône Delta and on the Don and Dnieper.

European Beaver

Scaly-tails

This isolated family is somewhat reminiscent of the flying squirrels in having (in most species) a flight membrane; the latter is however not stretched out by a bony spur from the hand joint, but by a cartilage spur from the elbow joint. There are two rows of large horny scales, arranged like tiles, on the underside of the base of the tail; the free edges of these scales grip the trunks of trees and thus help the animals when they are climbing. Scarcely any other mammals are so completely specialized for life in the tree-tops. There are about ten species.

Red-backed Scaly-tail, *Anomalurus erythronotus,* 11 + 9½ in. Found in the tropical forests of the Congo.

Red-backed Scaly-tail

are excellent climbers. They nearly all hibernate.

Common Dormouse, *Muscardinus avellanarius,* $2\frac{3}{4}+2\frac{1}{2}$ in. The soft fur is rusty yellow-brown on the back and paler on the belly. The neck is thicker than in ordinary mice, but the hairs on it are shorter than in the other dormice. They live in scrub and open woodland and are excellent climbers; they can hang by their back legs from the thinnest branches and can climb upside down. They make burrows or runs in the ground, but in the breeding season they build a spherical nest of grass and leaves, most often in thick undergrowth. They have 2-7 young in a litter and sometimes breed twice in the year. Dormice

Jumping Hares

A family with only a single species, possibly related to the scaly-tails.

Jumping Hare or **Spring Haas,** *Pedetes caffer,* $17\frac{1}{2}+15\frac{1}{2}$ in. Found on the plains and deserts of southern Africa, sometimes in very large colonies. The fur is rust-yellow on the back and whitish on the belly, and the tip of the tail is black. The front legs are short and the back legs long, the tail is long and bushy, and the eyes and ears are large. Jumping hares dig extensive burrows in the earth, using both the teeth and the front legs. They can jump 6-9 ft. and in emergency even more. They feed on leaves, roots and seeds.

Dormice

These animals have a very small thumb, a long and often bushy tail and they

Jumping Hare or Spring Haas (back), *and Lesser Egyptian Jerboa*

feed on hazel-nuts, mast, buds, berries and seeds. They start to hibernate in October, and during this period their body temperature is only a little above that of their surroundings. This is a southern species with its northern limit in Sweden. Elsewhere it lives in most parts of Europe, including Britain, southwards to the Pyrenees and south Italy.

Garden Dormouse, *Eliomys quercinus,* 5+4 in. The back is yellowish red-brown and there is a black area round the eye; the tail, of which only the outer half is bushy, is black on the upper side and white on the underside. Garden dormice are found in central and southern Europe and northwards to Finland. They live mostly in deciduous forests and in gardens and parks. They are mainly vegetarian, but like the other European rodents, they also eat insects, spiders, birds, eggs and small mammals. They set up their winter quarters in hollow trees or holes in the ground, which they line with leaves, grass and moss. They hibernate from September to April, and the female gives birth to 4-6 naked and blind young at the beginning of the summer.

Russian Dormouse, *Dyromys nitedula,* 4+3 in. These are very similar in habits and appearance to garden dormice; they are found in south-east Europe, the Caucasus and Asia Minor.

Fat Dormouse, *Glis glis,* 6½+5½ in. The back is ash-grey, the sides grey-brown and the belly white; there is a black patch round the eye. Fat dormice are found in central and southern Europe and western Asia. They are mainly

Common Dormouse (above), *and Garden Dormouse*

Fat Dormouse (above), *and Russian Dormouse*

nocturnal and feed chiefly on beech mast and hazel-nuts, but will also eat any animal they can kill. In autumn they collect stores which they place in their lined nests in the ground, in hollow trees or in deserted birds' nests. They begin to hibernate in September-October and do not wake up again before the end of April. The female gives birth to 3-7 naked and blind young at the beginning of June. Fat dormice have been introduced into Buckinghamshire and have now spread to neighbouring areas.

Jerboas

The jerboas have enormously elongated hind legs on which they hop rather than walk. The head and eyes are large, the snout is short with long whiskers, and the neck very short.

Lesser Egyptian Jerboa, *Jaculus jaculus,* 6¾ + 8½ in. The fur on the back is sandy yellow, and the belly is white. Lesser Egyptian jerboas occur in the Sahara as well as in the rest of North Africa, and in Arabia, Syria and Palestine, often living in large colonies. They are nocturnal vegetarians.

The *Birch Mouse* is usually classified close to the jerboas. The back is yellow-brown with a narrow black stripe running along the spine. The long tail can be used as a prehensile organ when the animal is climbing in trees. Birch mice go into hibernation in an underground nest in the autumn. They are found in Norway, Sweden, Finland, Denmark, Germany, Poland and the Carpathians, but not in Britain.

Mole-Rats

The blind mole-rats are excellent diggers and spend almost the whole of their lives underground. The limbs are short, and the fore-limbs in particular are used to dig with; the snout and head are also used for digging, and the powerful gnawing teeth serve to break up the earth. The eyes are hidden under the skin and the ears are very small. They feed mainly on roots.

European Mole-rat, *Spalax microphthalmus,* 7¾ in. Found in southern Russia and South Poland where they do damage in the fields by eating roots and making large mounds of earth, which are even larger than those of the common mole. There is a similar species in the Balkans and others in Africa and western Asia.

Naked Mole-rat, *Heterocephalus glaber,* 4 + 1½ in. These are very similar

Naked Mole-rat (back), *and European Mole-rat*

in habits to the other mole-rats. Apart from a few sparse hairs they are almost completely naked. The tiny eyes are rudimentary and almost covered by the thick eyelids. They dig long burrows in the earth and almost never come up to the surface. They are found in Somaliland, Abyssinia and the neighbouring parts of East Africa.

European Hamster (left) *and Golden Hamster*

The True Mice

This family is the richest in species of all the families of mammals. They may be classified into several subfamilies, including the hamsters, the voles and the rats and mice.

The *Hamsters* have short limbs, powerful incisor teeth and large cheek pouches.

European Hamster, *Cricetus cricetus,* 9¾ + 2¼ in. The back is yellow-brown, the belly black and there are white areas around the mouth and on the cheeks, elbows, knees and feet. European hamsters are found in western Siberia and the Caucasus and in most parts of central Europe. Each animal has a summer den, 20-25 in. below the surface of the ground, and the winter quarters may be twice as far down. A narrow main tunnel leads down to the nest, and there are several short runs going off at right angles, as well as places for reserve stores. They feed on grass, seeds and fruits and sometimes take insects, lizards, small birds and small mammals. Hamsters live alone for most of their lives; as soon as mating is over the female drives the male out of the burrow, and after 20 days' gestation gives birth

usually to 8-12 naked and blind young. There are two such litters in the year.

Golden Hamster, *Mesocricetus auratus,* 6¼ + 2 in. Golden hamsters are found wild in Syria but in recent years have been introduced as pets into many other countries. The period of gestation is only 10-14 days and each litter has 8-10 young, which are fully-grown and capable of breeding at an age of 3-4 months.

The *Voles* are small rodents with a short body, a broad head, and a very short tail. The have a mainly northern distribution.

Mole Vole, *Ellobius talpinus,* 3¾ + ½ in. Externally these animals resemble the lemming, but they are a separate species. The back is blackish-brown, the belly rust-yellow and the incisor teeth are very powerful and used for digging. They are found in southern Russia and central Asia.

Mole Vole (above); *Norway Lemming*
(centre); *North American Lemming* (below)

tunnels in the snow. The 2-11 young
are born in April-May and there are
several litters later during the summer.
In certain years their rate of breeding
is extraordinarily high and these "lemming years" are perhaps caused by particularly favourable climatic conditions.
These colossal increases in population
in the course of one or two years result
in mass migrations, during which the
lemmings ignore all obstacles, whether
mountain, lake or river, but keep right
on in the same direction, until finally
they are either drowned in the sea or
in lakes, or eaten by predators.

North American Lemming, *Lemmus
trimucronatus,* 5 + 1¼ in. Similar in appearance and habits to the Norway
lemming; found in North America, particularly around Hudson Bay.

In the *Collared Lemmings* of northern Siberia, Greenland and arctic North
America, the grey-brown summer coat
has a red-brown stripe on the neck.

Norway Lemming, *Lemmus lemmus,*
6 + ½ in. The only mammal which is
found only in Scandinavia. The back is
rust-brown, the belly yellowish-white
and the top of the head black. Norway lemmings go about more at night
than by day, and feed on grass, leaves,
heather, bark, moss, lichens, fungi and
carrion. They dig underground tunnels
and always run to shelter if chased, but
if they cannot escape in this way they
sit up on their hind legs in an aggressive attitude and hiss. In winter they dig

Musk Rat

Musk Rat, *Faber zibethicus,* 11¾ + 9¾ in. The coat is dark brown on the back and paler on the belly. The scaly tail is laterally compressed and the hind feet are webbed. Musk rats live along the banks of rivers and lakes and feed chiefly on marsh and water-plants. They are native to North America, but in 1905 three pairs were imported into Bohemia. These bred and spread themselves so successfully that they soon became a pest. Musk rats are known in the fur trade as musquash.

Water Vole

Water Vole, *Arvicola terrestris,* 7½ + 3½ in. The coat is shiny dark brown. Water voles are about the size of a rat and live mostly near to water in marshes and meadows, and in canals, dams and lakes, where they dig long tunnels in the earth, usually close beneath the surface. From these runs they push up small mounds of earth which resemble molehills. They swim and dive well, and are often out during the day searching for corn, clover, bark, and other plant food. They can do much damage by gnawing at the roots of small trees below the surface of the water; they store large winter reserves of corn and potatoes. A water vole may produce several litters in the course of the summer, each with 2-9 young. They are found over almost the whole of Europe and far into Asia.

Field Vole, *Microtus agrestis,* 4½ + 1½ in. This is a common mammal within its range, but is not often seen, since it lives most of the time in extensive runs among the stems of grasses. In winter field voles dig tunnels in the snow and line them with grass. They keep mostly to fields, marshes, heaths and hedges, but in the autumn they often go into woods

and do damage by barking young trees down near the roots. They feed principally on grass, roots, corn and leaves. Except during the winter months they may produce 4-10 young at intervals of only 3 weeks, that is about 10 litters per year, but on the other hand each animal does not live for more than 1½ years. Like other small rodents they have many enemies, amongst others, dogs, cats, foxes, badgers, stoats, weasels, owls, buzzards and other birds of prey. They are widespread over most of Europe, including England and Scotland.

Continental Field Vole, *Microtus arvalis,* 4½ + 1¼ in. Found in central Europe and far into Asia, but not in Britain, and very similar to the field vole, both in appearance and habits.

Bank Vole, *Clethrionomys glareolus,* 4 + 2 in. Found over the greater part of Europe and western Asia and common in most parts of Britain. Bank voles are active both by day and night, and feed on nuts, mast, corn and grass; in winter

(from above) *Field Vole, Continental Field Vole, Bank Vole, House Mouse*

they gnaw bark. They live on the edges of woods and in hedges and are remarkably good climbers. There are usually 3-4 litters a year, each with 2-6 young.

The *Mice* and *Rats* have a rather long snout with long whiskers, large round eyes, large naked ears and a long scaly and almost naked tail.

House Mouse, *Mus musculus,* $3\frac{1}{2} + 3\frac{1}{2}$ in. House mice live particularly in association with man, and are mostly found in lofts, stables, barns and other outhouses; during summer, however, they often move out into fields and gardens, and in autumn they are common in corn-ricks. House mice run fast and climb well, often in the most inaccessible places. They are omnivorous and feed amongst other things on corn, cheese, bread, sugar, and other foods, but may also make do with paper, soap and candles. They can enter cupboards, chests and furniture where they build large nests of chewed-up paper and straw. They produce a litter of 4-8 young every month or every alternate month throughout the year, but they live for only a year and a half. House mice originally came from Asia but they have now spread over almost the whole of the globe. *White mice* are domesticate albino forms of the house mouse.

Black Rat, *Rattus rattus,* $6\frac{1}{2} + 7\frac{1}{2}$ in. These may be recognized by the large, thin ears and the long tail which is longer than the body. The coat is usually black, but there is also a grey-brown race, the so-called Alexandrine rat. At one time the black rat was the commonest rat in Europe, but during the last 100-200 years it has been largely, although not completely, displaced by the brown rat, which is omnivorous and thrives better in a damp coastal climate. Black rats are better adapted to a continental climate; they feed almost exclusively on plant food such as corn, fruits and roots. They climb remarkably well and often live in the upper storeys of buildings and in roofs. The female produces several litters a year, each with 4-8 young. Black rats came originally from south-east Asia, but they are now found over almost the whole world. In Britain they are rather rare and are found mainly in some of the larger sea ports, which they have reached in foreign ships.

Brown Rat, *Rattus norvegicus,* 9½ + 8 in. The back is grey-brown and the belly whitish-grey. Although the legs are short and the feet small, brown rats can run fast and climb well. They use the incisor teeth to gnaw through wooden floors and planking, and the powerful claws to dig extensive underground runs, which have several emergency exits in addition to the entrance. Here and there in the runs there are wide parts which are used as eating-places or at nest sites. Brown rats normally come out at dusk, but where they are undisturbed they may also be seen around during the day. They swim and dive well and this has often led to their being confused with water voles. They are omnivorous and do millions of pounds worth of damage every year by eating corn, potatoes, turnips, and all kinds of foodstuffs; they also kill small animals such as young sparrows and chickens, and in lakes and ponds they destroy ducklings by pulling them down under water and drowning them before eating them. Food poisoning and plague and other infectious diseases can also be spread by brown rats. Man's attempts to control this efficient and hardy mammal are based partly on methods of preventing their access to food and shelter and partly on the use of traps, poison, guns or dogs. The rate of reproduction is astonishingly fast; they can produce up to 6-7 litters a year, each with 6-8 young, and they are capable of breeding when only 2-3 months old. On the other hand each rat lives only for about 2½ years. Brown

Brown Rat (above), *and Black Rat*

rats are considered to have come originally from Asia, whence they have spread to the farthest corners of the earth, often taken by ships, and in many places they have become an agricultural pest.

Common Field Mouse, *Apodemus sylvaticus,* 3½-4 + 4 in. Recognizable by the large eyes and ears and the long tail and hind legs. Field mice keep mostly to scrubland and fields. They feed principally on nuts, corn and cone seeds but may also eat insects. Under piles of branches and tree stumps or down in the burrows one often finds cones which field mice have gnawed. Unlike the squirrel they are unable to flay off the lowest scales of the cone, but have to gnaw them off one by one, and so a mouse-gnawed cone does not have a point at the base, but is rounded off below, and on the whole the job appears to be done with more care. They produce several litters in the year, each with 3-9 young; they may live for 4 years. They are found over most of Europe and far into Asia and are common in Britain.

The *Yellow-necked Field Mouse* is very similar to the common field mouse, but is a little larger and has a yellow band on the neck in front of the fore limbs. Its range is mainly in central Europe, and it is common in some areas of southern Britain.

Striped Field Mouse, *Apodemus agrarius,* 4+3¼ in. The coat is red-brown with a black stripe along the back. Striped field mice live in underground runs, in fields, parks, scrubland and on the edges of woods. They are often seen out during the day and, like the harvest mouse, can use the tail as a prehensile organ. They feed on insects, worms, roots and seeds. They produce 3-4 litters during the course of the summer, each with 4-8 young, and may live for 3 years. Striped field mice are found in central Europe from the Rhine eastwards to Siberia, but not in Britain.

Harvest Mouse, *Micromys minutus,* 2½+2¼ in. Harvest mice are found mainly in open country, often in oat fields and among high grasses, where they can climb up and down the stalks, using the tail as a prehensile organ. They feed on seeds, berries, greenstuff and insects. They make their summer nest about 18 in. above the ground, between corn or wild grass stems and line it with grass leaves. The females produce 4-9 tiny young in each litter, and may have several litters in the year. In winter they live in runs in the ground or in barns. They live for 2-3 years. They are found in most parts of Europe and eastwards to the Pacific Ocean, and are quite common in some parts of southern Britain.

(from above) *Harvest Mouse, Common Field Mouse, Striped Field Mouse, House Mouse (field form)*

Old World Porcupines

Porcupines have very large and characteristic quills on the back which can be erected by powerful skin muscles. They include the short-tailed porcupines with species in Africa, tropical Asia and southern Italy, and the brush-tailed porcupines of Africa and southeast Asia.

Common Porcupine

Common Porcupine, *Hystrix cristata,* $25+4 \perp 9$ in. This is one of the short-tailed porcupines and it is found along the coasts of the Mediterranean in Spain, South Italy, Sicily and North Africa; another slightly larger form is found in tropical Africa. The body is blackish-brown with white hair on the back of the neck, and long spines which are ringed black and white. There is a mane along the neck made up of very long spines. Some of the spines on the back can reach a length of 15 in. and are indeed the longest in the whole of the animal kingdom. The spines can be raised and lowered by the powerful skin muscles, which cover must of the body; the spines on the tail are much shorter. When excited a porcupine rubs the spines against each other and produces a loud rustling sound.

Porcupines live almost always alone; by day they sleep in their underground run, and at night they move around slowly searching for food – mainly juicy plants, roots, fruits and bark. When eating they hold the food with the front paws – a common habit among the rodents. As they move about at night they rustle the spines continuously, apparently as a warning signal comparable with the rattle of a rattlesnake. The mating season is in the spring, and after a gestation period of two months the female produces 2-4 young in her warm hole down in the ground.

New World Porcupines

These are arboreal animals with sharp claws and rather short quills. The group includes the tree-porcupines of Central and South America, which have prehensile tails, and the North American porcupines, which have rather short flat tails.

Brazilian Tree Porcupine, *Coendou prehensilis,* $25+17\frac{1}{2}$ in. Found in forest

country from northernmost South America through the Amazon region to Bolivia. The tail is long and prehensile; unlike the spider monkeys and the kinkajous which curl the tail round a branch from above downwards, tree porcupines coil it from beneath the branch and then up around it. As an adaptation to this method of coiling, the upper side of the top of the tail is naked and covered with horny scales. If the animal wants to crawl from one branch to another, it takes hold with the back feet and the tail and then stretches the body straight out at an angle and tries to get hold of the new branch with its front feet. If it succeeds it releases its hold on the old branch. A dog that is bold enough to attack a tree porcupine gets a mouthful of sharp spines, which penetrate deeper and deeper the more it tries to get loose; the dog is unable to close its mouth, and if it does not get help, it soon chokes.

Hairy Tree Porcupine, *Coendou villosus,* 25+17 in. Found in the forests of central and South Brazil and Paraguay. The spines are hidden by long woolly hair. Hairy tree porcupines spend their whole life in the tree-tops, where they sleep by day rolled up in the angle of a branch, and climb around slowly at night searching for leaves and fruits. The female produces one or sometimes two young in October.

Canadian Porcupine, *Erethizon dorsatum,* 23+7 in. Found in forest areas from Labrador southwards to the mountains of Pennsylvania and Virginia. The coat consists partly of shortish soft hairs and partly of long stiff bristles. In ad-

(from above) *Brazilian Tree Porcupine, Hairy Tree Porcupine, Canadian Porcupine*

dition there are long stout spines running along from the top of the head to the tip of the tail, but these are mostly hidden by the long hairs. The colour is blackish-brown with white areas on the hairs and spines. When climbing a tree they grip the trunk with all four limbs, stretch to full length, then move the back legs up close to the front ones and stretch themselves forwards again, rather in the manner of a looper caterpillar; at the same time they press the flat muscular tail in against the trunk. During the summer they feed on greenstuff, but in winter they eat bark almost exclusively, mostly of elm, poplar and conifers, and thus do much damage. The spines are loosely fixed and can easily be released from the skin to pierce an enemy. The female produces one or sometimes 2-4 young in a hollow tree in April-May.

Coypu

Coypus

Coypu, *Myocastor coypus,* 17 + 17 in. Found in the southern part of South America, living along lakes and rivers where there are plenty of water plants. The head is large, the ears small and the legs short with webbed toes; the long tail is hairless and scaly. The chestnut-brown fur is long and thick and consists of soft woolly hairs, which are completely hidden by long guard hairs. The incisor teeth are large with reddish-yellow enamel on the front. On land coypus move about rather slowly, but they swim with great ease, using the front feet and the tail. They often live in colonies and dig holes in river banks; their food consists of grass, leaves, corn and roots. After a gestation period of 2-3 months the female gives birth to 5-10 young; there are usually two such litters in the year. The 4-5 nipples are positioned near to the sides of the body so that the young can suckle whilst the mother is in the water. The young can manage in the water from the time of birth, and they are themselves capable of breeding when they are 8 months old. As a rule coypus live for 4-5 years. In the fur trade the pelt of the coypu is always known as "nutria". Only the woolly part of the pelt is used; the coarse guard hairs are removed, leaving the soft grey-blue nutria fur. Coypus have been introduced as fur-farm animals into several countries; unfortunately they sometimes escape from these farms and do damage to river banks.

Peruvian Chinchilla (above), *and Chile Chinchilla*

Mountain Chinchilla (above), *and Viscacha*

Chinchillas

The chinchillas resemble rabbits and hares in having very long back legs, but they differ in having a long bushy tail and only 3 or 4 toes on the feet. The very thick fur has been used for centuries by the South American Indians, and has been much exported to Europe in the last 150 years.

Peruvian Chinchilla, *Chinchilla chinchilla,* 11 + 10 in. Found in the western parts of the Andes, being most frequent in Peru, but probably exterminated in the wild state. Chinchillas feed exclusively on plant food, and have several litters in the year, each with 4-6 young. They are farmed for their pelts in America and elsewhere.

Chile Chinchilla, *Chinchilla velligera,* 9½ + 4½ in. Found in northern Chile, where it is now protected. The fur is known in the trade as "Bastard Chinchilla".

Mountain Chinchilla, *Lagidium peruanum,* 15 + 9½ in. Found living gregariously at altitudes of 9,000-15,000 ft. in Ecuador, Peru and Bolivia.

Viscacha, *Lagostomus maximus,* 19 + 7 in. Found on the pampas of Argentina, living gregariously. The back is mousegrey and the belly white; the head is large with large eyes and ears. Viscachas undermine the earth with their tunnels and holes; one often finds colonies consisting of 12-15 holes with 20-40 animals. These underground burrows are a danger to horses, and for this reason, and also on account of their attractive soft fur, they are much hunted.

(above, left) *Paca,* (right) *Sooty Agouti;* (below, left) *Pacarana,* (right) *Golden Agouti*

Guinea-Pig-like Rodents

This group, which is found wild only in South America, includes the world's largest rodent. The claws are very broad and rather like small hooves, and the tail is short or completely rudimentary.

Paca, *Coelogenys paca,* 27 in. Found along the banks of dense marshes in Central and South America. The back is red-brown and the belly yellowish, and there are 3-4 long rows of white stripes on the sides. Pacas spend the day in underground holes and come out at night to search for food, often entering sugar plantations and maize fields. When scared and unable to reach home they try to escape by swimming and diving.

Pacarana, *Dinomys branickii,* 28+12 in. Found in the mountain forests of north-western South America. The coat is dark brown with two longitudinal rows of white spots on the sides; the tail is quite long and hairy. Pacaranas are nocturnal and usually live in forest areas; very little is known about them. Some authorities consider that the pacarana should be placed in a separate family.

Golden Agouti, *Dasyprocta aguti,* 15+ ½ in. Found in the forests of Brazil, Guiana and Venezuela. The fur is smooth and reddish-gold in colour with a black-brown sheen. Golden agoutis feed on fruits, birds and eggs.

Sooty Agouti, *Dasyprocta fuliginosa.* Found in Ecuador and Brazil. The fur is black with a silvery sheen.

Aperea, *Cavia aperea,* 7½ in. Found in southern Brazil and Paraguay, living

Guinea-Pigs. (above) *Angora Guinea-Pig and 'Bolivian' or Smooth Cavy;* (below) *Wild Guinea-Pig and 'Abyssinian' Cavy*

in holes and underground tunnels. The domestic guinea-pig is derived from the related but smaller Peruvian Cavy, *Cavia cutleri.*

Domestic Guinea-Pig, *Cavia porcellus.* The fur of guinea-pigs is very variable in texture and colour; there are several races, such as the **Abyssinian,** and the **Bolivian** or **Smooth Cavy.** They live exclusively on plant food. The gestation period is about 63 days and there may be 2-3 litters (or more) per year, each with 2-3 young.

Capybara, *Hydrochoerus capybara,* 39 ⊥ 19 in. This is the largest rodent in the world. Capybaras live in small groups in marshy areas along the rivers of South America, and feed on water plants, grass, maize, rice and sugar cane. The gestation period is about 5½ months and there are usually two young in the litter.

The *Patagonian Cavy,* 18+2 in., is a harelike rodent, found on the plains of Patagonia. The long hind legs each have three toes with hoof-like claws.

Capybara

Edentates

A T one time the armadillos, anteaters and sloths were grouped together with the pangolins and aardvark in the order Edentata—meaning toothless—because they all showed a great reduction in their dentition. Nowadays the pangolins and the aardvark are separated off in two separate orders, leaving the armadillos, anteaters and sloths to form the order Edentata. At first sight these three types of animal may seem rather far apart from each other, but a study of extinct fossil forms has shown that they had common ancestors in the past. They all walk on the soles of the feet, have powerful claws, the third finger is strongly developed and the teeth, if present, have no enamel. All the living Edentates occur exclusively in Central and South America.

Armadillos

In these peculiar mammals the upper side of the body is covered with plates, which consist partly of horn produced by the outer layer of the skin, and partly of bone from the underlying layers. The plates are arranged regularly in transverse bands over the back, limbs and tail. The armadillos are the only mammals which have a true integumentary skeleton – a skeleton in the skin – and they also show the greatest variation in the number of teeth. The number of teeth may, in fact, vary from 28-100, all of which are weak and ill adapted for biting and chewing. The tongue is worm-shaped and well fitted for licking up ants and termites. Armadillos feed principally on insects but also eat snails, larvae and worms and sometimes also plant food. They are found from Texas and Mexico southwards right to Patagonia, mostly in dry open country and as a rule avoiding dense forest areas.

Six-banded Armadillo, *Euphractus sexcinctus,* 16+8 in. Found in Brazil, Guiana, Bolivia, Paraguay and the Argentine. Six-banded armadillos have no special hunting-place but wander about during the night, and, in places with plenty of insect life, they dig down about six feet into the ground, and remain in these burrows during the day. As a rule the female gives birth to only one young at a time.

The *Giant Armadillo* is more than twice the size of the six-banded arma-

Six-banded Armadillo

dillo; it is found in the forests of South America, from Guiana to Argentina.

The *Three-banded Armadillo* can roll itself up into a ball; the claws are very long and only their tips touch the ground when the animal is walking.

The *Fairy Armadillo* from western

Great Anteater

Argentina and Bolivia is only about the size of a mole and lives entirely underground. As might be expected the eyes and ears are very small.

Anteaters

The anteaters are surely the most specialized insect-eaters amongst the mammals, especially in their ability to dig up and eat termites and ants. They are completely toothless and the front part of the skull is drawn out into a tube with a mouth opening at the end which is so small that it allows room only for the 20 in. long sticky worm-like tongue to be slipped in and out. The front limbs have very powerful digging claws, with which they tear down termite colonies to get at the inmates.

Great Anteater, *Myrmecophaga tridactyla*, 51 + 27 in. Found in the Argentine, Paraguay, Brazil, Ecuador and north to Guatemala. The dark coat has long coarse hairs, especially on the back and tail, but the elongated head has short hairs. There is a white-edged black band running from the front of the neck over the shoulders. The front paws are turned inwards, so that the animal walks on the outer sides of the strong claws. When asleep a great anteater lays its bushy tail over the body to form a kind of roof. The food consists almost exclusively of termites and ants. In spring the female bears a single young, which she carries about on her back.

Tamandua, *Tamandua tetradactyla*, 24 + 15 in. A nocturnal animal living in forest country from Mexico to the Argentine and Paraguay, and feeding mainly on ants and termites. The yellowish-white coat has short hairs and is marked with black on the flanks. The tail is long and prehensile.

Sloths

Sloths live in the upper branches of trees, nearly always climbing and hanging upside down. The long coarse coat looks rather like matted grass, and usually has algae growing in it, which gives it a greenish tinge; in addition the fur is nearly always infested with certain small moths, which are found only on sloths. The head is broad, the jaw short and the ears are hidden in the fur. The front legs are longer than the back legs, and the toes are united in a common skin, so that only the long curved claws are free. The tail is quite short and serves as a flap over the rump. The large brown molars lack enamel and are the only teeth present. Nearly all other mammals have 7 neck vertebrae but in sloths the number varies between 6 and 9. Sloths are nocturnal animals which feed on leaf-buds and fruits. They bear only a single young at a time, which the mother carries about on her belly. If a sloth comes down to the ground it is almost helpless and can only move about very slowly and unsteadily, supporting itself on the forearms.

Two-toed Sloth, *Choloepus didactylus,* 27 in. Found in the tropical forests of northern South America. The coat is olive-brown, and the snout and the soles of the feet are naked. There are two claws on the front limbs and three on the back legs. The long tongue resembles that of an anteater.

The *Three-toed Sloth* has nine neck vertebrae and three claws on each limb. It is found in Brazil, Bolivia and the Argentine.

Tamandua

Two-toed Sloth

Pangolins

THIS order contains only a single genus with several species, which are found in Africa, south-east Asia and Indonesia. Apart from the sides of the head and the belly which are hairy, the whole animal is covered with large, sharp-edged, overlapping horny plates; the digging claws are very powerful. The sticky tongue is very long and extensible, but there are no teeth. Pangolins are able to roll themselves up into a ball as a form of protection.

Temminck's Pangolin, *Manis temmincki,* 29+29 in. Found on grassy plains in East and South Africa. The plates are yellowish-brown in colour. Temminck's pangolins feed at night on ants and termites. The female has only a single young in each year.

Temminck's Pangolin

Aardvarks

THIS order contains a single genus, with only one species.

Cape Aardvark, *Orycteropus afer,* 39 +31 in. The head is elongated, the front part forming a trunk-like snout ending in a small mouth with a long sticky thread-like tongue. The teeth lack enamel, have no roots and continue to grow throughout life. The back is strongly arched, the root of the tail is very powerful, and the four claws on the front feet are adapted for digging. The ears are about six inches long, and the coat long-haired and yellowish-grey. Aardvarks spend the day in a hole in the earth and go out at night in search of ant and termite colonies, which they break open with their claws to get at the insects inside. They are found in Africa south of the Sahara.

Cape-Aardvark

Bats

BATS are the only mammals which are able to fly actively, as opposed to just gliding. There are a great number of species, grouped into fourteen families, and they form the second largest order of mammals—only the rodents have more species. Feeding conditions seem to be uncommonly favourable for bats, both for the relatively few large Fruit Bats and for the numerous small Insectivorous Bats. The small bats are never likely to starve for they only have to catch insects for a short time before their stomachs are full. Many of them sleep the whole day and fly out at sunset to hunt for an hour; then they sleep the rest of the night, until sunrise, when they again hunt for a short time. In spite of these relatively short periods of feeding the bats are second only to the Insectivores in the amount they eat, relative to their body weight. The wings consist of thin almost naked skin, stretched between the limbs, body and tail. This flight membrane is principally stretched between the much elongated bones of the lower arm, hand and fingers. The thumb, however, is not elongated; it has a claw and is free from the flight membrane, and sometimes there is also a claw on the index finger. Bats sleep by day, hanging upside down by the curved claws of the back limbs; they do not usually fly out until dusk. They spend their sleeping time in hollow trees, church towers, lofts and other sheltered places, as a rule crowded together, often in hundreds, sometimes even in thousands. When flying they find their way by their sharp sense of hearing; they produce ultrasonic sounds, which are received back as echoes and serve to show the position of obstacles, rather on the principle of echo-sounding. In temperate Europe bats hibernate for 4-6 months, during which time their temperature falls with that of their surroundings and may even drop below freezing point. The body temperature also falls during their ordinary periods of sleep. Bats pair up before hibernation, but mating does not take place until the spring and the young are born in May-July. As a rule the female has only one young (at the most two) which is unable to fly immediately; she flies around with it on her belly during the time that she is feeding it.

Fruit Bats

The fruit bats have a long snout, claws on both the thumb and index fingers, and the tail is free from the flight membrane or it may be absent. They feed

Malay Fruit Bat or Flying Fox, and Indian Flying Fox

on fruits and honey-rich flowers, and are found in the tropics and subtropics, except in America.

Malay Fruit Bat, or **Flying Fox,** *Pteropus vampyrus,* 15 in. Found in the East Indies. This is the largest of all bats, with a wing-span of 59 in. The fur is brown and the wings are black. By day fruit-bats hang in large parties in their roosting trees with both body and head wrapped up in the glossy wings; in the evening they fly out in search of fruits and may do much damage in banana and fig plantations. When eating, they hang on by one hind leg and carry the food to the mouth with the other; sometimes they also catch insects.

Indian Flying Fox, *Pteropus giganteus,*

12 in. Very similar in appearance and habits to the Malay fruit-bat, but somewhat smaller in size. Indian flying foxes occur in India, Burma, and Ceylon. There are several other species.

Insectivorous Bats

The insectivorous bats have a short snout, very small eyes and very large ears. The teeth are pointed and well adapted for seizing and eating insects, on which most of them live exclusively; some however also feed on fruits, fish, blood and flesh. Insectivorous bats are found almost everywhere in tropical and temperate regions and there are several hundred species, of which about 16 are found in Europe, and of these 13 occur in Britain.

Vampire Bat, *Vampyrus spectrum,* $6\frac{1}{4}$ in. Found in the tropical forests of the Guianas and Brazil. Vampires belong among the leaf-nosed bats, which have lobes of skin on the snout. They have no tail and the wing-span is about 27

Vampire Bat

in. At one time it was thought that they could kill a man by sucking his blood whilst he was asleep. It is now known that they feed chiefly on insects and fruits. The term vampire would have been more appropriately applied to the true blood-sucking bats (see below).

Indian Vampire Bat, *Megaderma lyra,* 3½ + 1 in. Found in India and Ceylon. The large ears are united in the middle line. Indian vampires feed mainly on frogs, and insects, but also take small birds.

Great Blood-sucking Bat, *Desmodus rotundus,* 2¾ in. Found from Mexico to Brazil and Paraguay. This and two other related species are the true blood-sucking bats. The outer incisors and canines of the upper jaw are very sharp and are used to pierce the skin of sleeping horses, cattle and men. They do not actually suck the blood from the wound, but rather lap it up with their long tongues as it flows out.

Tomb Bat (above), *and Naked Bat*

Tomb Bat, *Rhinopoma microphyllum,* 2¼ + 2¼ in. Found in large numbers in Egypt, Arabia and Persia. This is one of the long-tailed bats. The body is pale grey and there is a flat swelling on the snout; the flight membrane does not reach back to the long thin tail. Tomb bats feed on insects.

Naked Bat, *Cheiromeles torquatus,* 5½ + 2 in. Found in Malaya, Sumatra, Java and Borneo. The body is almost naked; the skin is thick and folded, and on the breast it is developed into a deep pouch which contains the nipples. The young remains in this pouch during the flights of the mother.

The *Horseshoe Bats* are so called from the horseshoe-shaped folds of skin around the nostrils.

Lesser Horseshoe Bat, *Rhinolophus hipposideros* 1½ + 1¼ in.; wing-span 9 in. Found in central and southern

Indian Vampire Bat (above), *and Great Blood-sucking Bat*

Europe, including Britain, North Africa and south-west Asia. Lesser horseshoe bats feed mainly on flies and other small insects. They hibernate until late in the spring.

Great Horseshoe Bat, *Rhinolophus fer-rum-equinum,* 2½ + 1½ in.; wing span 13 in. Found in central and southern Europe, Africa and southern Asia eastwards to Japan, with a distinct sub-species in England. Greater horseshoe bats feed mostly on dung-beetles, cockchafers and moths, which they catch in flight.

Noctule Bat, *Nyctalus noctula,* 3 + 2 in.; wing span 15 in. Found almost everywhere in Europe and central Asia. Noctules fly very high and fast and swoop down on large insects such as cockchafers and dung-beetles. They shelter in hollow trees, often in large numbers, and come out early in the evening. They go on flying until late in the autumn and hibernate in ruins and old buildings.

Barbastelle Bat, *Barbastella barbastel-lus,* 2 + 2 in.; wing span 10½ in. Widespread in central and southern Europe, North Africa and south-west Asia. Barbastelle bats have proportionately short wings, but fly very dexterously. They come out early in the evening from hollow trees, roofs and holes in walls to hunt for mayflies, dragonflies and moths. In winter they hibernate down in cellars and come out again in early spring.

(from above) *Lesser Horseshoe Bat, Greater Horseshoe Bat, Noctule Bat, Barbastelle Bat, Serotine Bat, Long-eared Bat, Common Pipistrelle*

Serotine Bat, *Eptesicus serotinus,* 2¾ + 2 in.; wing span 10½ in. Found in Europe as far north as England and Denmark, and eastwards into Asia. Serotine bats fly low and come out later in the evening than most bats.

Northern Serotine Bat, *Eptesicus nilssoni*, 2¾ + 1¾ in.; wing span 10½ in. Found in Scandinavia and in the mountain regions of central Europe.

Long-eared Bat, *Plecotus auritus*, 2 + 1½ in.; wing span 9½ in. Found throughout Europe, eastwards into Asia and southwards to North Africa. The ears are almost as long as the body and proportionally longer than in any other animal; they can be bent backwards like a pair of broad horns, and when the animal is at rest they are usually tucked in under the wings. Long-eared bats catch insects in flight and also hover with whirring wings and snap them off the leaves of trees. They come out late in the evening and may hunt the whole night. By day they shelter in towers, lofts and outhouses. They hibernate in hollow trees and cellars.

Common Pipistrelle, *Pipistrellus pipistrellus*, 1¾ + 1¼ in.; wing span 8 in. Found from Ireland and Spain to eastern Asia, with its southern limit in the Mediterranean and Transcaucasia. This is the smallest European bat. It frequents the neighbourhood of houses and usually comes out while it is still light, to hunt for gnats and moths. By day it hangs hidden under bark, in hollow trees, under roofs and in church towers.

Parti-coloured Bat, *Vespertilio murinus*, 2½ + 1½ in.; wing span 11½ in. Found from Scandinavia to the Alps and central Asia. The fur is dark brown, but the hairs have grey tips so that the

(from above) *Northern Serotine Bat, Parti-coloured Bat, Bechstein's Bat, Natterer's Bat, Daubenton's Bat, Mouse-eared Bat, Whiskered Bat*

general appearance is mould-grey. Parti-coloured bats live mainly in towns and come out after dark to hunt for moths and beetles. They continue to hunt until January-February and then hibernate in the cracks and fissures of old build-ings, often in large numbers together, and usually hanging horizontally.

Bechstein's Bat, *Myotis bechsteini,* 2 + 1½ in.; wing span 11 in. Found in cen-tral Europe and westwards to England. Bechstein's bats emerge from hiber-nation late in the spring. They have only rarely been found in southern Eng-land.

Natterer's Bat, *Myotis nattereri,* 2 + 1½ in.; wing span 10 in. Found from Ire-land and Spain through central Europe and far into Asia. The back edge of the wing has a fringe of short stiff hairs. Natterer's bats are usually seen in wood-land areas, flying very low and rather slowly. They shelter by day in hollow trees and holes in walls.

Daubenton's or **Water Bat,** *Myotis dau-bentoni,* 1¾ + 1½ in.; wing span 9½ in. Found in Europe and northern Asia. Daubenton's bats are seen particularly near fresh water, where they fly close down over the water surface. They shel-ter by day in hollow trees and lofts, sometimes in large groups; they often hibernate in cellars.

Mouse-eared Bat, *Myotis myotis,* 3¼ + 2 in.; wing span 15 in. Found in central Europe; there is a closely related form in the Mediterranean region. Mouse-eared bats are often seen in towns, where they rest under roofs or in cel-lars, often in groups of 30 to 50 together.

Plantain or Butterfly Bat (above), *and Weltwitsch's Bat*

They may eat smaller bats. They fly low and rather clumsily.

Whiskered Bat, *Myotis mystacinus,* 1¾ + 1¼ in. Found over most of Europe as far west as Ireland, and also in Asia. Whiskered bats are so named after the prominent hairs on the upper lip, which are, however, only slightly longer than in certain other bats. They come out of hibernation in April and may some-times be seen hunting in the middle of the day. They often hibernate in cellars.

Plantain or **Butterfly Bat,** *Kerivoula picta,* 2¼ + 2 in. A strikingly mottled bat, more brightly coloured than is usual among bats. It occurs in Ceylon and India. It spends the day hanging in banana plants.

Weltwitsch's Bat, *Myotis weltwitschi,* 3¼ + 1⅛ in. A small bat from West Africa which has orange markings on the flight membrane, particularly along the digits.

Insectivores

THE Insectivores are small short-legged mammals, with a snout which is drawn out into a trunk. Most of them have large toes with sharp claws on both the front and back legs, and they walk on the soles of the feet. The greater part of their food consists of insects, larvae and worms; many of them consume an enormous amount of food relative to their size. Most Insectivores are nocturnal, probably not because they cannot find food by day, but because they are better protected in the dark. The Insectivore families are all rather different from each other, and it is difficult to name many characters which they have in common. The teeth are, however, fairly characteristic; they are adapted primarily for piercing and dividing the food, and not for chewing it up into small pieces. The molars are pointed, and the whole dentition acts rather like a pair of pincers with a mass of points. The canine teeth are not particularly prominent. The relatively large number of teeth and their lack of specialization, and the poor development of the brain, suggest that the Insectivores are a primitive group of mammals. Hearing and smell are well developed, but the eyes are usually small and their sight is weak. The main area of distribution of the Insectivores is in the temperate zone of the northern hemisphere; there are none in South America or Australia. Many of the Insectivore families have a very restricted distribution.

Tenrecs

These have spines in the fur, at any rate in the young; the snout, head and body are more elongated than in the hedgehog. The family is found in Madagascar and some of the neighbouring islands.

Tenrec, *Centetes ecaudatus,* 15½ in. This is the largest Insectivore and, unlike others, it has large canines in the lower jaw. The brownish coat consists of hairs and coarse bristles, with short spines on the top of the head and neck; the young also have spines along the back. Tenrecs dig burrows in the ground, and are mainly nocturnal. They are hunted for their tasty flesh. At the beginning of February the female gives birth to 12-20 young in the underground burrow, and thus produces more young per litter than any other mammal. Tenrecs live in the forest-clad mountain regions of Madagascar, and feed on insects, worms, snails, lizards and fruit.

Ericulus setosus, 7¾+¾ in., (illustrated), is another of the tenrecs, in which the spines extend over the whole of the back.

Otter-Shrews

Otter-Shrew, *Potamogale velox,* 12+ 11½ in. The only species in its family. The snout is broad and the tail is laterally compressed and thick at the root. The toes are not webbed, but otter-shrews can swim quite fast. They are found in streams and rivers in West Africa and feed on fish and crustaceans.

Tenrec (above); *Ericulus Tenrec* (below)

Almiqui or Cuba-Solenodon (above);
Agouta or Haiti Solenodon (below)

The Solenodons

This family contains only two species, which live in the West Indies. The snout is long and drawn out to form a trunk, the tail is long and scaly, and there are strong claws, particularly on the front legs. They are rare nocturnal animals.

Otter-shrew

Agouta or **Haiti Solenodon,** *Solenodon paradoxus,* 11 + 9½ in. The long trunk is used for searching for insects and larvae. They are found in Haiti, where they are much hunted by introduced mongooses.

Almiqui or **Cuba Solenodon,** *Solenodon cubanus,* 11 + 9½ in. A closely related form found in Cuba.

Moles

This family contains several species with tiny eyes and ears usually hidden in the fur. The *Star-nosed Mole* from the north-eastern United States and southeast Canada has a star-shaped growth on the snout. The *Russian Desman,* has a long laterally compressed tail and webbed toes; it lives in south-east Europe and western Asia.

European Mole, *Talpa europaea,* 5½ + 1 in. The snout is short and movable. There are no outer ears, and the eyes are smaller than pin-heads and usually covered with fur. The greater part of the front limbs is hidden in the body and only the broad digging-paws with strong claws stick out. Moles dig long burrows in the earth; molehills consist of loose earth which has been pushed up by the head and shoulders during these digging operations. The breeding nest is made under a particularly large mound, and the 2-6 naked and blind young are born in it in May-June. They live mainly on earthworms and collect living reserves of these by biting the front end of the worm so that they cannot escape from the runs. Moles are hunted by carnivores, birds of prey, and herons. In winter they dig down deeper into the earth, but do not hibernate. The are widespread over Europe and Asia, and are common in most areas of Britain, but do not occur in Ireland.

Shrews

The shrews are very small animals with long, pointed snout and soft fur. The two middle front teeth in the lower jaw are turned forward, and in some the tips of the teeth are coloured red-brown.

(from above) *Common Shrew, Alpine Shrew White-toothed Shrew*

European Mole

Common Shrew, *Sorex araneus,* 2¾ + 1½ in. Found especially in meadows and near to water. In May-June the female has 6-8 tiny, naked and blind young, which can look after themselves when they are about 4 weeks old. They usually breed twice in the summer following their birth and then die in the autumn. Their food consists mainly of insects, spiders, snails and worms. Carnivores sometimes bite and kill them, but then leave them uneaten because they have an unpleasant musky smell; nevertheless owls eat many shrews. Common shrews are widely distributed throughout Europe and northern Asia and are common in Britain, although absent from Ireland.

Alpine Shrew, *Sorex alpinus,* 2¾ + 2¾ in. This is a woodland shrew, found in the Pyrenees and the Alps right up above the tree-belt. The tips of the teeth are red.

Pigmy Shrew, *Sorex minutus,* 2¼ + 1½ in. It occurs in northern and central

(from above) *Pigmy Shrew, Pachyura,*
Water Shrew

Europe and Asia, and is not uncommon in Britain; it is the only shrew found in Ireland.

Water Shrew, *Neomys fodiens,* 4+2½ in. Water shrews may be recognized by the fringe of stiff hairs extending along the edge of the feet and on the underside of the tail. They live in mouse and mole runs, or they may dig their own burrows in loose earth. The nest is made in a hollow tree root or in a hole in the ground. Water shrews usually breed twice a year, each litter having 4-8 small, naked and blind young. The food consists of insects, worms, snails, bivalve molluscs, crustaceans and small fish, but they may also take frogs and mice. They are widespread over central and southern Europe and southern Siberia, and are common in many places in Britain.

White-toothed Shrew, *Crocidura russula,* 3+1½ in. One of a number of related species in which the tips of the teeth are white. It often comes into houses, stables and barns. It occurs in central and southern Europe, northern Africa and parts of Asia.

Pachyura etrusca, 1½+1 in., (illustrated), the smallest mammal in Europe, if not in the world, is found in gardens and houses in South Europe.

Hedgehogs

The hedgehogs form a family of Insectivores found only in the Old World. There are several species in Europe, Asia and North Africa.

Common Hedgehog, *Erinaceus europaeus,* 11+1 in. The back and sides of the body are covered with numerous stiff, ¾ inch-long horny spines. The claws are strong and well adapted for digging. Hedgehogs live in small woods, parks, scrubland and gardens. By day they lie up in sheltered places. As night falls they come out to find insects, worms and snails or fallen fruits. They track their food by smell and by sound; their sense of hearing is good. Now and again

Hedgehog

they may attack snakes, frogs and mice, and exceptionally may take eggs and chickens. At the beginning of the summer the sow hedgehog bears 5-6 blind, whitish-grey young with short soft spines; they get their adult dress in the course of a few weeks. The young suckle for a month and then go out with the sow to learn how to find food, before moving off on their own. In autumn hedgehogs collect moss and withered leaves in a lair under a tree root or in a stone wall, where they hibernate; during this period the body temperature may sink to 42°F. They come out of hibernation in March-April. Common hedgehogs are found (in two forms) over the whole of Europe except in the northern parts of Norway, Sweden, Finland and Russia. The boundary between the West European and East European forms is on the River Oder. They are found almost everywhere in Britain and Ireland.

Elephant-Shrews

The elephant-shrews have a trunk-like snout and long thin back legs.

Four-toed Elephant-Shrew, *Petrodromus sultani,* 6¾ + 5 in. These are found in Africa; they hop about rather like miniature kangaroos.

North African Elephant-Shrew, *Elephantulus rozeti,* 5½ + 4¼ in. Found in the Atlas Mountains in North Africa, where they search for insects early in the morning.

Tree-Shrews

This family has a number of species living in India and south-east Asia.

Four-toed Elephant-Shrew (above), and North African Elephant-Shrew

They have so many characters in common with the lower Primates that many zoologists now consider that they should be placed in that order, rather than among the Insectivores.

Tana, *Tupaia tana,* 7¼ + 7¾ in. Found in Sumatra and Borneo. They may reach high up into the hills but are also seen in cultivated areas and even inside houses. They feed by day on insects and fruit.

Tana

Flying Lemurs

THIS order contains one or two closely related genera and there has been considerable controversy about the place it should have in zoological classification. They have been classified at various times among the lower Primates and the Bats. They are now usually placed in a separate order close to the Bats and Insectivores.

The most striking feature character of the flying lemurs is the large parachute-like flying membrane, known as a *patagium,* which is more strongly developed than in the flying squirrels or flying marsupials. The lower arm and lower leg are much lengthened, and all the fingers and toes are strongly developed. The flying membrane is hairy on both sides and starts at the neck, joins up all the limbs right to the claws and also takes in the tail. Flying lemurs climb in trees in search of fruits and leaves and they can jump from the tree-tops and glide for about 200 ft. They are found in Malaya, Borneo, Java, Sumatra and the Philippines.

Colugo or Flying Lemur

Colugo or **Flying Lemur,** *Galeopithecus volans,* 19½ + 4 in. Found in the tropical forests of the Philippines. They are nocturnal and sleep by day suspended upside-down by the claws, with the head and tail tucked in between the legs. The short red-brown fur has the same colour as tree bark. At dusk they begin to climb around in search of food. They seldom come down from the trees and can only move slowly and with difficulty on the ground. The litter consists of a single naked young, which the female carries about on her belly for a long time.

Marsupials

IN all the mammals described so far, that is the placental mammals, the embryo is maintained in its early stages in the uterus, or womb, of the pregnant mother, and is retained there until it has reached a relatively advanced stage of development. With the marsupials we come to a group of mammals in which the period spent in the uterus of the mother is relatively short and the young animal is born in a very undeveloped condition. Immediately after it is born the young marsupial crawls into the pouch, or marsupium, which is situated on the belly of the mother; in a few marsupials the pouch is lacking. The pouch itself is a fold of the belly skin surrounding the nipples; in this pocket of skin the tiny young get protection, warmth and food. In the large kangaroos the gestation period is only 39 days, in the opossum only 13 days; but after the new-born young has reached the pouch it will remain there, fastened to the nipples, for as long as 65-70 days. In the pouch the young clings on to a nipple, which swells up into its mouth, so that it can remain hanging. A new-born marsupial is so poorly developed that it cannot itself suckle, and the milk is squirted into its mouth by the contraction of muscles in the nipples of the mother. The larynx of the young is so arranged that the milk from the mother can run in through a kind of tube. Even after the youngster has eventually started to suckle itself it still remains for a long time in the mother's pouch; and when it does come out several months later it still uses the pouch as its home for quite a long time. The female seldom protects or defends her young, even though she has them with her for a longer period than in other mammals. Marsupials are found mainly in Australia but there are also several in New Guinea and they extend north-westwards as far as Celebes; there are also two families in America.

It is interesting to note the range of adaptation among the marsupials. Just as we have found that different groups of placental mammals are adapted for various modes of life, so among the marsupials there are species adapted for carnivorous, insectivorous or vegetarian diets, or for digging, swimming, jumping or even aerial gliding. The living marsupials are grouped into eight families. The species illustrated here belong to two sub-orders: the *polyprotodonts,* which are flesh- and insect-eaters, and include the opossums, the native cats and the bandicoots, and the *diprotodonts,* which are vegetarian, and include the wombats, phalangers and kangaroos. The polyprotodonts have a large number of front teeth; 6-10 in the upper jaw and 6-8 in the lower jaw. The diprotodonts have only 2-6 front teeth in the upper jaw and two front teeth in the lower jaw.

Opossums

The opossums are rat-like marsupials with pointed heads and very numerous teeth; there are no fewer than 10 incisors in the upper jaw and 8 in the lower jaw. The tail is usually long and is commonly developed as a prehensile organ. Some of them have a large number of nipples, and in such cases the pouch is often lacking and the young merely shelter among the thick fur on the belly and fasten themselves to the nipples. The family is found exclusively in America.

North American Opossum, *Didelphys virginiana,* $20\frac{1}{2}+12$ in. Opossums are found in dense woodland in the eastern part of the United States, where they climb about in the trees with great skill, using their long toes and powerful prehensile tail. They are mainly nocturnal and feed on small mammals, birds, eggs, frogs, insects and worms, and also on vegetables and fruits. Sometimes they

Thick-tailed Opossum

run amok in hen-houses, killing whole flocks. When threatened they appear to feign death–hence the expression to play 'possum. This is often regarded as a trick, but in fact it is probably some kind of nervous shock, for the animal lies down, almost without breathing, and when it does begin to move again it will often make several fruitless attempts to rise before it is successful. Opossums are much sought after for their fur, and millions of their skins come on to the market every year.

Thick-tailed Opossum, *Metachirus crassicaudatus,* $10\frac{1}{2}+7\frac{3}{4}$ in. The fur is red-brown and the area at the root of the tail is thick and hairy. They are excellent climbers, and, as in many other opossums, the larger young are carried on the mother's back with their tails curling round hers. Thick-tailed opossums live on the forest-clad river islands of the Guianas, Brazil and Argentina, and feed on birds and small mammals. They sleep in hollow trees or in nests which they build themselves.

North American Opossum

Water Opossum

about the whole day. It is very blood-thirsty and will attack any birds or mammals of its own size or even larger.

The *Caenolestids* or *Selvas* form a group of shrew-like marsupials, now classified in a sub-order on their own. They are found only in South America, and they have no pouch.

Carnivorous Marsupials (Dasyurids and Banded Anteater)

This group corresponds not only to the Carnivores but also to the Insectivores among the placental mammals. They have a total of 42-46 teeth, made up of small incisors, large canines and pointed molars, so that their dentition is considerably more specialized than that of the opossums. They also differ from the latter in having a hairy tail, which cannot be used as a prehensile organ, and in having only four toes on the back legs. They are found in Australia, Tasmania, New Guinea and neighbouring islands.

Water Opossum, *Chironectes minimus,* 15+15 in. This is the only marsupial which is adapted for an aquatic life; it has webs between the toes of the back feet. The upper parts are ash-grey with a black stripe running along the spine and broad black areas stretching down the sides; the belly is white. Water opossums feed on fish and other aquatic animals, but although carnivorous they have large cheek pouches, which are otherwise characteristic of certain purely vegetarian mammals. Another peculiarity which is, however, found in many other marsupials, is that the pouch has its opening facing backwards; this might seem to be an adaptation to swimming if it were not for the fact that the female remains on land for as long as she has her five young in the pouch. Water opossums are found from Guatemala to southern Brazil.

Short-tailed Opossum, *Peramys americana,* 5¾+2 in. Found in the forest areas along the rivers of eastern Brazil. This is a lively animal which wanders

Short-tailed Opossum

Brush-tailed-Phascogale

Fat-tailed Pouched Mouse, *Sminthopsis crassicaudata,* 4+3 in. Found in Australia, where there are several other related species. In this species the tail is thick at the base. The female has 8-10 nipples.

Jerboa-like Pouched Mouse, *Antechino-mys laniger,* 3½+4¾ in. Found in south-western Queensland and western New South Wales. The jerboa-like pouched mice have large ears, a long tufted tail and a pointed snout, and are very similar in appearance to both the elephant-shrews and the jerboas, and like them they hop about on the back legs, which have very long feet. They have a pouch only during the breeding season. They feed on insects.

Brush-tailed Phascogale, *Phascogale penicillata,* 10+9 in. Small marsupials looking rather like squirrels, with soft, woolly grey fur, and a tail which has long hairs on the outer half. They are nocturnal and sleep by day in hollow trees, where they also have their young. They hunt among the branches for insects and other small animals. They are found almost everywhere in Australia.

The *Native Cats* have dark fur with white spots and a very hairy tail. The pouch, which is only developed in females which have young, usually has an opening in the middle and six nipples.

Common Native Cat, *Dasyurus viverrinus,* 17¾+9¾ in. Found in the moun-

Fat-tailed Pouched Mouse

Jerboa-like Pouched Mouse

Common Native Cat

tain forests of South Australia and in Tasmania. Native cats feed on small mammals, birds and insects. The young are not more than one-third of an inch long at birth, which is tiny even for a new-born marsupial. The young cling fast to the nipples for more than two months and only start to look after themselves when they are 4½ months old.

The *Tasmanian Devil* has coal-black fur, usually with a white marking on the front of the chest, and is found in the inaccessible forest country of Tasmania. It feeds almost exclusively on other marsupials and on birds.

Tasmanian Wolf, *Thylacinus cynosephalus,* 51 + 19 in. This is the largest of all the carnivorous marsupials. It looks like a large dog, but the tail is thicker at the base, as is so often the case in marsupials, and the mouth opening is much larger. The coat is greyish-brown with

black cross-stripes on the back. Tasmanian wolves were at one time widespread in Tasmania, but were much hunted because they attacked the settlers' sheep and poultry. They also killed kangaroos, bandicoots, echidnas and other smaller animals. Nowadays they are found only in the western mountain area of Tasmania and are extremely rare. They are nocturnal and can see only with difficulty in sunshine. There are usually four young in a litter.

Banded Anteater, *Myrmecobius fasciatus,* 10 + 7 in. Found only in certain areas of southern Australia, having been exterminated throughout the greater part of their original range, probably because they are slow, defenceless and rather trusting. The fur is yellowish with broad black cross-bands running over the back, the head is brownish with a black stripe running through each eye and the tail is long and bushy. There

Tasmanian Wolf

are about 50 teeth in the dentition, which is more than in any other marsupial. Banded anteaters live chiefly in forest areas where there are hollow trees in which they can hide. They feed mainly on termites which they suck up with the long, narrow and sticky tongue. The female has four nipples, but no pouch, and the four young have to make do with what protection they can find among the hairs on the mother's belly.

Marsupial Mole

Marsupial Moles

Marsupial Mole, *Notoryctes typhlops*, 4¾ + ¾ in. A digging marsupial, in which the four inner toes of all the feet have enormously powerful digging claws. There is a hard horny shield on the upper part of the snout, which serves as a ram in digging. The eyes are completely rudimentary and overgrown by muscle and skin. The tail is short and leathery and marked with rings. The pouch opens backwards and has two nipples. The soft silky fur is pale yellow with a golden sheen. Marsupial moles seem to dig principally for beetles and for the larvae of ants, moths and butterflies, which are found among the roots of acacia trees. They live in the sandy regions of South Australia and were first observed in 1888. They often come up to the surface after rain, even during the day. They evidently do most of their digging just below the surface and do not excavate deep burrows.

Banded Anteater

Bandicoots

The bandicoots are digging marsupials with elongated back legs; the small second and third toes are united whilst the fourth and fifth toes are free and strongly developed. The head is narrow and the snout is long and pointed. The pouch opens backwards.

Western Rabbit Bandicoot, *Thalacomys lagotis*, 15¾ + 10 in. Found in Western Australia and now very rare. The coat

which have only two functional toes. The fur is grey-brown in colour. Pig-footed bandicoots feed on insects and tree bark. They were once found over almost the whole of central Australia, but are now very rare.

Native Bears and Wombats

This group contains a number of stout, heavily-built marsupials.

Koala, or **Native Bear,** *Phascolarctos cinereus,* 24 in. Koalas are entirely arboreal. On the front feet the first two digits can be opposed to the other three, giving a kind of pincer action. The head is very thick, the ears are long-haired, and the black-brown snout is naked almost up to the eyes. The thick soft fur is reddish-grey on the back and whitish-yellow on the belly. Koalas are found

Western Rabbit Bandicoot (above), *and Pig-footed Bandicoot*

is greyish, and the tail is grey at the base, black in the middle and white at the end. Rabbits bandicoots are more nocturnal than most of the other marsupials; they feed on insects, larvae, worms and roots.

Pig-footed Bandicoot, *Choeropus castanotis,* 9¾ + 4 in. About the size of a small rabbit, but with long hind legs,

Koala or Native Bear

Wombat

in eastern Australia but they have been much hunted for their beautiful thick fur and are now extinct in many areas. They move about mostly at night, and feed almost exclusively on eucalyptus leaves. The female usually has only one young at a time, which remains in the pouch for three months; she then carries it about on her back for a further six months.

The *Wombats* are burrowing marsupials, of which there are four or five species in eastern Australia and Tasmania, all very similar to each other. Their front teeth are very like those of the rodents.

Wombat, *Phascolomys mitchelli,* 39 in. Found in the dense forests of south-eastern Australia. they feed exclusively on plant food. Wombats use their short

powerful claws to dig long burrows, in which they spend the day and where they also bear their young. The pouch opens backwards, as in the other digging marsupials.

Phalangerines

Climbing marsupials, in which all the fingers and toes have claws, except the big toe which is opposable and thus well adapted for gripping the boughs of trees. The tail is long and often prehensile. Phalangerines are distributed from Celebes to Tasmania, and include the flying phalangers, cuscus, dormouse-phalangers, honey mouse and Australian opossums.

Short-headed Flying Phalanger, *Petaurus breviceps,* 7+8 in. A very large flight membrane runs from the outer side of the little finger to the base of the hind foot. The thick soft fur is ash-grey, the snout short and the eyes large. Flying phalangers feed on leaves, flower-buds,

Short-headed Flying Phalanger

Common Dormause-Phalanger

honey and insects and are found throughout eastern Australia. The young remain in the pouch only while they are naked, and then they crawl out first on to the belly of the mother, and later perch themselves on her back.

Common Dormouse-Phalanger, *Dromicia nana,* $4^3/_4 + 3^1/_8$ in. These tiny animals are usually fast asleep during the day, but at night they are very lively and climb around looking for honey and insects. The tongue is well adapted for sucking honey out of flowers, for it is split at the tip to form a little brush. In the winter months they go into a kind of hibernation, curling themselves up into a ball and living on the fat reserves which are stored in the root of the tail. They are found in Tasmania and south-eastern Australia; there are other species in New Guinea and in other parts of Australia.

The *Cuscuses* are cat-sized, nocturnal marsupials found in the area between Celebes and Northern Queensland. The outermost part of the prehensile tail is naked on both the upper and lower sides.

Spotted Cuscus, *Phalanger maculatus,* $23^1/_2 + 19^1/_2$ in. Found in New Guinea and northern Australia. The thick, soft, silky fur is very variable in colour; the upper side is white, yellowish or greyish, marked with large irregular rust-red, brown or black spots, and the belly is white. The face is rust-yellow in the young, and almost pure yellow in the older animals; completely white specimens are also found. As a rule the female is more uniform in colour than the male. Spotted cuscuses feed on the leaves and fruits of trees and may also take some animal food. They are slow-moving animals, almost as lethargic as sloths. The female usually has only one young at a time.

Closely related to the cuscus are the *Australian Opossums,* which, however, have hairy tails and large ears. They are arboreal animals, found in Australia and Tasmania; their soft fur is marketed

Spotted Cuscus, female and male

Short-eared Opossum

smaller animals. The female gives birth usually to one young at a time, which is carried in the pouch for two months and then for a further period on her back.

Short-eared Opossum, *Trichosurus caninus,* 23 + 17 in. Found in the bushy country of eastern Australia. This species has sohter ears than the common phalanger and occurs in a dark and a pale form.

Honey Mouse, *Tarsipes rostratus,* 2¾ + 3¼ in. Found in the southern parts of Western Australia. A small marsupial with a long pointed snout, a thin worm-like tongue and a long prehensile tail; the teeth and claws are small and weakly developed. At night honey mice climb about actively in the trees, searching for honey and insects.

under the name of "Australian Opossum".

The *Common Phalanger* or *Long-eared Opossum* has large ears and thick fur, which varies in colour from grey-brown through rust-red to yellow ochre. Long-eared opossums move about slowly in the tree-tops at night and are among the most common mammals in Australia. They feed principally on the leaves of gum trees, but also take birds and

Kangaroos

The kangaroo group contains a considerable number of species, which vary in size from the rabbit-sized tree-kangaroos through the wallabies to the gigantic true kangaroos, which are as large as a grown man. The ears are long, and the head has something both hare-like and deer-like about it. The front part of the body is relatively slender and the short and weak front limbs are used principally for gripping. The back part of the body, on the other hand, is very powerful and so are the long hind legs and the long muscular tail. The small front legs have five fingers with relatively small claws, whilst the hind legs have very long feet which lack the inner toes; of the other four toes the two outermost ones are very strongly devel-

Honey Mouse

Yellow-footed Rock-Wallaby

The *Rock-Wallabies* have naked noses and long hairy tails which do not taper towards the tip. The central claw of the hind feet is very short. They occur in the rocky regions of Australia but are not found in Tasmania or New Guinea.

Yellow-footed Rock-Wallaby, *Petrogale xanthopus,* $25\frac{1}{2} + 25\frac{1}{2}$ in. Found in the mountain regions of eastern Australia. The fur is greyish on the back and white on the belly, and the tail has alternate yellow and brown-black rings. Rock-wallabies are active and agile

Red Kangaroo, male (back), *and female*

oped and have large almost hooflike claws; the remaining two small toes are united and can be turned in at right angles to the rest of the foot. Apart from some of the smaller species the kangaroos are diurnal animals, and all are vegetarian, most of them grazing on the dry grass plains. When a kangaroo runs away it hops on its hind legs and uses the tail as a rudder. They can jump to heights of 9 ft. and lengths of 30 ft., a feat unsurpassed by any other mammal except the impala. Hearing is the best developed of the senses, and the ears are always moving; sight is not particularly good, and the sense of smell is poor. During the breeding season the males fight savagely amongst themselves, rising up on their tails and attacking with the sharp claws of the back legs. Kangaroos usually have only one young at a time.

climbers, and can even get up steep cliffs. Although nocturnal they will often come out during the day to sun themselves.

The *True Kangaroos* and *Wallabies* have a more self-coloured coat than the small and medium-sized species, and the typical kangaroo characters of long back legs and a long powerful tail are very pronounced. These has been some doubt about how many species there are, but there are probably about 5 true kangaroos in addition to about 18 wallabies. They are found over the whole of the Australian continent and Tasmania.

Red Kangaroo, *Marcropus rufus,* 75+ 64 in. This is the largest of all the kangaroos and also the largest marsupial. When sitting up on its hind legs and tail an old male is as tall as a grown man. The fur is peculiar in that it consists only of woolly hairs without any guard hairs. The male is red with a grey head, pale tail, pale legs and black tips to the toes. The female, which is considerably smaller, is blue-grey with a pale belly. This description of the colour applies mainly to the animals in eastern Australia, for there is much variation between the geographical races. Thus in Western Australia there are red females, and in other places both sexes are grey-blue. The red colour is said to be mainly due to a secretion from the skin. The gestation period is 39 days and the new-born young is only $1\frac{1}{8}$ in. long. When 6 months old it begins to poke its head out of the pouch and at 8 months is starts to leave the pouch and wander about on iits own, but it

White-throated Tree-Kangaroo

goes on using the pouch as a home for a considerable time.

The *Tree-Kangaroos* from New Guinea and north Queensland spend their lives in trees, and have large, pointed claws, which help them in climbing. The fore limbs are proportionately longer and the hind limbs shorter than in the jumping kangaroos. The long tail is flexible but is not used as a prehensile organ or as a support. They feed on leaves, buds and fruits.

White-throated Tree-Kangaroo, *Dendrolagus leucogenys,* 29¾ + 26 in. Found in New Guinea. This is only one of several species of tree-kangaroo.

Monotremes

THE Monotremes differ considerably from all the other mammals. First, they lay eggs with leathery shells and secondly, they have a cloaca, that is a single opening for the alimentary canal and the urinary tract, as in the birds and the poikilothermal, or cold-blooded, vertebrates. In addition they lack nipples, so that when the young want to suckle, the female has to lie on her back and let the milk trickle out on to two patches of skin on the belly. The milk is yellowish and creamy and rich in albumen. Monotremes have a much lower body temperature than the other mammals, and in fact their temperature changes with that of the environment and may vary nearly as much as that of the cold-blooded reptiles. In several respects Monotremes do indeed stand closer to the reptiles than to any of the other mammals. The males have a spur on the heel, connected with a gland which is thought to produce a secretion which acts as an excitant during the breeding season. The order contains only two families, the echidnas or spiny anteaters and the duck-billed platypus.

Echidnas

The echidnas have a tubular snout and a small toothless mouth with a long, wormlike, sticky tongue. The body is covered with hairs and spines, the tail is short and the digging claws are long and powerful, particularly on the hind legs where they face outwards. Echidnas live mainly on larvae, ants and termites, which they pick up with their sticky tongues. They lay a single egg, about half an inch long, which is placed into a pouch on the belly, where it is incubated. When ready to hatch, the young one makes a hole in the horny shell with a small egg-tooth situated on the snout; it is about 3-3½ in. long at the time of hatching. Echidnas are nocturnal animals with well-developed hearing and smell. There are four or five species found in New Guinea, eastern Australia and Tasmania.

The *Australian Echidna* has numerous long spines on the back and a relatively short snout. It moves about slowly at night, but is a fast digger. When threatened it can dig down into the soil in the course of a few minutes, or it may roll itself into a ball with the spines sticking out, like a hedgehog, but it has no other form of defence. The body temperature fluctuates between 80°F and 90°F, depending on the temperature of the air.

Long-beaked Echidna, *Zaglossus bruijni,* 19 in. In this species, which comes from New Guinea, the snout is about

Long-beaked Echidna

the toes webbed. Platypuses live along the banks of rivers and lakes and feed mainly on bivalve mollucs, but will also take snails, worms and crustaceans. They dig long burrows–several yards in length–in the banks, and line them with soft plants; the entrance opens under water. Platypuses can remain submerged for about ten minutes. Unlike the echidnas the female has no pouch, but lays her two eggs inside the burrow and incubates them until they hatch. The young remain in the burrow until they are about 4½ in. long, but they do not venture into the water with the female until they are about 8 in. long.

Platypuses are active both by day and night. They are found in south-east Australia and Tasmania.

twice as long as the head, and the limbs are considerably longer than those of the Australian echidna. There are several races.

Platypuses

The family contains only a single species.

Duck-billed Platypus, *Ornithorhynchus anatinus,* 17½ + 5½ in. The platypus has a flat, duck-like bill covered with thin, naked skin; there are teeth in the young stages, but these are soon replaced by flat horny plates. At the base of the horny bill there is a plate which protects the eyes during digging. The thick brown fur consists of long close-packed guard hairs and soft woolly hairs. The tail is flat, the claws powerful and

Duck-billed Platypus

Index

Animals marked with an asterisk are illustrated in the text.
This index gives only the popular names of the animals.

Index

Index